The Mind

The Mind

Consciousness, Prediction, and the Brain

E. Bruce Goldstein

The MIT Press
Cambridge, Massachusetts
London, England

© 2020 The Massachusetts Institute of Technology

All rights reserved. No part of this book may be reproduced in any form by any electronic or mechanical means (including photocopying, recording, or information storage and retrieval) without permission in writing from the publisher.

This book was set in Stone Serif and Stone Sans by Westchester Publishing Services. Printed and bound in the United States of America.

Library of Congress Cataloging-in-Publication Data

Names: Goldstein, E. Bruce, 1941– author.
Title: The mind : consciousness, prediction, and the brain /
 E. Bruce Goldstein.
Description: Cambridge, Massachusetts : MIT Press, [2019] |
 Includes bibliographical references and index.
Identifiers: LCCN 2019042164 | ISBN 9780262044066 (hardcover)
Subjects: LCSH: Cognition. | Brain. | Thought and thinking.
Classification: LCC BF311 .G5828 2019 | DDC 150—dc23
LC record available at https://lccn.loc.gov/2019042164

10 9 8 7 6 5 4 3 2 1

Contents

Preface   vii
Acknowledgments   xi

1 **Introduction to the Mind**   1
2 **Consciousness and Experience**   31
3 **The Hidden Mind**   65
4 **The Predictive Mind I: Perceiving and Acting**   93
5 **The Predictive Mind II: Language, Music, Memory, and Social Prediction**   121
6 **Dynamic Highways of the Mind**   155

Notes   183
Further Reading   219
Name Index   223
Subject Index   227

Preface

Here's a problem. Write a short book—under fifty thousand words—titled *The Mind*. What do you include? This crux of this problem is that the mind is responsible for everything we experience, as well as lots of things we are unaware of that create our experience. It is about our perceptions, physical movements, reading and listening, thoughts, emotions, and interactions with others, among other things. So at the outset let's establish that it is impossible to cover the mind in fifty thousand words, or, for that matter, even in multiple full-length volumes. Nonetheless, I thought I'd give it a try.

My starting point was a course I was teaching for the Osher Lifetime Learning Institute (OLLI) at the University of Pittsburgh, the University of Arizona, and Carnegie-Mellon University titled "Your Amazing Mind." My OLLI course begins with three questions: What is the mind? What is consciousness? and What happens in the mind that we are not conscious of? This discussion leads down many interesting byways (What is it like to be conscious? Are non-human animals conscious? Can we know what someone else is experiencing? Is it possible that making a decision involves an unconscious process?) and inevitably to the brain, which not only creates all of our experiences, but accomplishes that creation largely "behind the scenes," outside of our awareness.

After covering these topics in chapters 1–3, and with about twenty-five thousand words to go, I faced a decision: how to go beyond this basic introductory material to cover specific cognitive capacities like perception, memory, language, and everything else the mind does? Because I did not have the space to devote a separate chapter to each of these capacities, I decided to focus on a principle that holds across cognitive capacities.

That principle is the principle of prediction—figuring out what is going to happen next—and in chapters 4 and 5 I consider eye movements, visual object perception, tactile sensations, language, music, memory, and social interactions. What unfolds over these two chapters is the idea that there is a process that operates across many different capacities. But as I tell this story, it also becomes apparent that this process actually comprises many different processes. Predicting where the eye is going to move next and predicting what a person is going to do next involve an automatic mechanistic process in the case of the eye, and a much more cognitive process in the case of the person. What this means is that prediction has a lot to teach us about the mind in general. In addition, we can discuss prediction at both the behavioral level (how does prediction operate as we read?) and the physiological level (how is neural firing affected by how predictable a sentence is?).

Finally, the last chapter. What should it be about? In early drafts, I outlined chapters on mind wandering, thinking, and social interaction, each of which are important behaviors. But in the end, I looked back at the chapters about prediction and asked myself, "What do all the behaviors and physiological processes associated with prediction have in common?" One answer is that they all include communication between different places in the brain. Because these connections are so central to prediction, and to most other cognitive processes as well, it seemed like a good choice for the final chapter. After all, the mind, which is created by the brain, emerges not from the firing of neurons in one specialized area but from communications that travel across what could be called "highways of the mind."

Thus, although this book does not "cover the field," it does provide an introduction to age-old questions about the mind while also describing current research, some of which is moving so quickly that ten years from now, we may be looking back on this research as "the beginning of something big." So I invite you to engage in this book, which, although it is brief, will give you a taste of some of the exciting things we know and are finding out about the mind.

Acknowledgments

This book has benefited greatly from the contributions of the many people who helped me by reading and providing feedback on my various writing projects over the years, especially ten editions of *Sensation and Perception* (1980–2017) and five editions of *Cognitive Psychology* (2001–2019). For help with *The Mind*, I thank Tessa Warren, who awakened me to the importance of the cognitive process of prediction; Jessica Hana-Andrews, who guided me through the complexities of the default mode for an early draft of the book; and the anonymous reviewers for the MIT Press who provided valuable feedback on the prospectus and the final manuscript. I thank Marianne Taflinger, my editor for the early editions of *Cognitive Psychology*, for her untiring support of that project. Special thanks to Shannon LeMay-Finn, master developmental editor, who has provided advice and support on numerous editions of my textbooks and read and commented on all the chapters of this book, which reads more clearly because of her. I am also grateful to the Osher Lifelong Learning Institute (OLLI), which has afforded me the opportunity to teach courses on the mind to "returning students" (over fifty) at the University of Pittsburgh, Carnegie Mellon University, and the University of

Arizona. The material in the first three chapters of *The Mind* is based on courses I have taught at OLLI.

I especially thank my editor for *The Mind*, Phil Laughlin of the MIT Press, for being open to taking on this project and for his unwavering support throughout, and Judith Feldmann, who guided production from initial manuscript to final book. Thanks also to Alex Hoopes for obtaining all those permissions on short notice, and to Mary Reilly for her willingness to make the fixes that made the illustration program clearer.

Finally, I dedicate this book to two people who have played meaningful roles in my life: Ken King, who, as psychology editor at Wadsworth Publishing, "discovered" me in 1976, when I was an Assistant Professor at the University of Pittsburgh. He launched my textbook-writing career by convincing me to write *Sensation and Perception*, and then supported the publication of many of the editions that followed. And the last and most important dedication is to my wife, Barbara, who has been there for me through it all. Thank you, Barbara.

1 Introduction to the Mind

Sixteen years ago, a patient I will call Sam suffered injuries that put him into a coma from which he never woke up. He showed no signs of awareness or ability to communicate. Observing Sam, lying in the care facility, it wouldn't be unreasonable to conclude that "there's nobody in there." But is that true? Sam doesn't move or respond, but does that mean he doesn't have a mind? Is there any probability that his eyes, which appear to be vacantly staring into space, could be perceiving, and that these perceptions might be leading to thoughts?

These are the questions Lorina Naci and coworkers were asking when they placed Sam in a brain scanner that measured increases and decreases in electrical activity throughout his brain, and then showed him an eight-minute excerpt from an episode of *Alfred Hitchcock Presents* called "Bang! You're Dead."[1] In this TV program, a five-year-old boy finds his uncle's revolver, partially loads it with bullets, and begins playing with it in his room, making believe he is firing it—saying "bang, bang"—but not actually pulling the trigger.

The drama escalates when the boy enters a room where his parents are entertaining a number of people. He points the gun at people threateningly, saying "bang, bang," to pretend he is shooting. Will he pull the trigger and actually shoot? Will someone be killed? These questions race through most viewers' minds. (There was a reason Hitchcock was called "the master of suspense.") At the end of the film, the gun goes off, the bullet smashes a mirror, the boy's father grabs the gun, and the audience breathes a sigh of relief.

When this film was shown to healthy participants while they were in the scanner, changes in their brain activity were linked to what was happening in the movie. Activity was highest at suspenseful moments, such as when the child was pointing the gun at someone. So the viewer's brains were not just responding to the pattern of light and dark on the screen, or to the images on the screen; their brain activity was being driven by what they were seeing *and* by the movie's plot. And—here is the important point—to understand the plot, it was necessary to understand things that were not specifically presented in the movie, like why the gun is important (it is dangerous when loaded); what guns can do (they can kill people); and that the five-year-old boy was probably not aware of the danger that he could accidentally kill someone.

So how did Sam's brain respond to the movie? Amazingly, his response was the same as that of the healthy participants: activity increased during periods of tension and decreased when danger wasn't imminent. These findings indicate that Sam was not only seeing the images and hearing the soundtrack but reacting to the movie's plot. His brain activity therefore indicated that Sam was consciously aware, so "someone was in there."

This story about Sam, who appears to have a mental life despite appearances to the contrary, carries an important message as we embark on the adventure of understanding the mind. Perhaps the most important message is that the mind is hidden from view. Sam is an extreme case, because he can't move or talk, but you will see that the "normal" mind also holds many secrets. Just as we can't know exactly what Sam is experiencing, so we do not know exactly what other people are experiencing, even though they are able to tell us about their thoughts and observations.

And although you may be aware of your own thoughts and observations, you are unaware of most of what is happening in your mind. This means that as you understand what you are reading right now, hidden processes are operating within your mind, but beneath your awareness, that are making this understanding possible.

As you read this book, you will see how research has revealed many of these secret aspects of the mind's operation. This is no trivial thing, because your mind not only makes it possible for you to read this text and understand the plots of movies but also is responsible for who you are and what you do. It creates your thoughts, perceptions, desires, emotions, memories, language, and physical actions. It guides your decision-making and problem-solving. It has been compared to a computer, although your "brain computer" outperforms your smart phone, laptop, or even a powerful supercomputer on many tasks. And, of course, your mind does something else that computers can't even dream of (if only they could dream!): it creates your consciousness of what is out there, what is going on with your body, and, simply, what it is like to be you.

In this book, we will be exploring what the mind is, what it does, and how it does it. The first step is to look at some of the mind's achievements. As we do, we will see that the mind is not monolithic but multifaceted, involving multiple functions and mechanisms.

The Multifaceted Mind

One way to appreciate the multifaceted nature of the mind is to consider some of the ways "mind" can be used in a sentence. Here are a few possibilities:

1. He was able to call to mind what he was doing on the day of the accident. *The mind as involved in memory.*
2. If you put your mind to it, I'm sure you can solve the math problem. *The mind as problem solver.*
3. I haven't made up my mind yet *or* I'm of two minds about that. *The mind as decision maker.*
4. I know you well enough to read your mind *or* You read my mind. *The mind as involved in social interactions.*
5. He is of sound mind and body *or* He is out of his mind. *The mind as involved in mental health.*
6. A mind is a terrible thing to waste. *The mind as valuable.*
7. He has a beautiful mind. *Some people's minds are especially creative or exemplary.*[2]

These statements tell us some important things about what the mind does. Statements 1, 2, 3, and 4, which highlight the mind's role in memory, problem-solving, decision-making, and interacting with other people, are related to the following definition of the mind:

> *The mind creates and controls functions such as perception, attention, memory, emotions, language, deciding, thinking, and reasoning, as well as taking physical actions to achieve our goals.*

Statements 5, 6, and 7 emphasize the importance and amazing abilities of the mind. The mind is something related to our health, it is valuable, and we consider some people's minds extraordinary. But one of the messages of this book is that the idea of the mind as amazing is not reserved for "extraordinary" minds, because even the most routine things—recognizing a person, having a conversation, deciding what food to buy at the supermarket—become amazing when we consider the properties of the mind that enable us to achieve these familiar activities.

What exactly are the properties of the mind? What are its characteristics? How does it operate? A book called *The Mind* has a lot to explain. Not only does the mind do a lot, but just about everything the mind is asked to achieve turns out to be more complicated than it first appears. Take, for example, opening your eyes and seeing a scene before you. You might think seeing the scene can't be that complicated, because, after all, light reflected from the scene creates a picture of the scene on the retina that lines the back of your eye.

One reason for the difficulty of perception is that the picture of the scene on the retina is ambiguous. When the three-dimensional scene "out there" is represented by a picture on the flat surface of the retina, objects that are at different depths in the scene can appear right next to each other in the picture. If this isn't obvious from looking out at the scene around you, close one eye, hold up one of your fingers, and place it next to a far-away object in the scene. When this scene-with-superimposed-finger becomes the picture on the retina, the finger and object appear adjacent to each other, even though they are far apart in the scene. The mind solves this "adjacency problem," plus many others, so that we don't have to deal with them; we just open our eyes, and we see!

Another example of something we achieve easily in the face of great complexity is understanding language. The stimulus for language, like the stimulus for perception, can be ambiguous, with the same word having different meanings, depending on the contents and structure of the sentence it appears in. Consider the following two sentences:

1. Time flies like an arrow.
2. Fruit flies like a banana.

Many things are going on in these sentences, including different meanings for "flies" (1: flies = moves; 2: flies = a type of bug) and "like" (1: could replace "like" with "similar to"; 2: could replace "like" with "appreciate"). Even simple, apparently straightforward sentences are more complicated than they appear. When we read "a car flew off the bridge," we are pretty certain that the car doesn't have wings and isn't a special kind of flying machine. In fact, we would probably be right to guess that the car was involved in an accident, that it may have been submerged in water or smashed on the ground, that the bridge possibly sustained some damage, and that the driver was in danger of being injured or worse. All these conclusions from a simple six-word sentence!

As we describe how the mind tames the complexities of perception, language, and many other abilities, we will begin to appreciate that the mind is not simply an "identification machine" that catalogs objects and meanings but a sophisticated problem solver that uses mechanisms we are largely unconscious of. We will also see that these hidden mechanisms often use knowledge that we have accumulated about the world. We know, from past experience, that when a car "flies off a bridge," it is unlikely that it will actually fly.

As we begin exploring the complexity of the mind's operation, we will be looking at a number of the functions listed in our definition on page 4 but will narrow our focus by considering just some of the mind's functions. We will narrow our focus further by basing our discussion on discoveries made using the scientific approach, in which hypotheses about the mind are tested by making controlled observations or running experiments. The scientific approach to the study of the mind has had an interesting history, which one could describe, with a nod to the Beatles, as "A Long and Winding Road."

Approaches to the Mind

We begin our discussion of the scientific study of the mind by considering an experiment by Franciscus Donders, which highlights the fact that we can't directly measure the mind.

Donders's Pioneering Experiment

The first attempts to measure the mind in the scientific laboratory began in the mid-nineteenth century in the laboratory of Franciscus Donders (1818–1889), professor of physiology at the University of Utrecht in the Netherlands.[3] Donders's goal was to determine how long it took for a person to make a decision. He answered this question by measuring *reaction times* under two conditions. In the first condition, measuring *simple reaction time*, the participant sees a light flash on a screen and pushes a button as quickly as possible. The reaction time is the time between when the light appears and when the button is pushed. In the second condition, measuring *choice reaction time*, there are now two lights. The participant's task is to press the left button if the left light flashes and the right button if the right light flashes. This condition involves the decision: which button should I press?

Donders found that the choice reaction time was about one-tenth of a second longer than the simple reaction time, and so concluded that it takes one-tenth of a second to make a decision in this situation. But the real importance of this experiment is in Donders's reasoning, which we can appreciate by considering the diagrams in figure 1.1.

In both the simple condition (fig. 1.1a) and the choice condition (fig. 1.1b), the time between the stimulus (lights flashing) and the behavioral response (pushing a button) is measured. The mental response (seeing the light; deciding which button to push) is not measured. The fact that it took participants one-tenth of a second longer to respond in the choice reaction time condition caused Donders to *infer* that this was the extra "mental time" that it took to make a decision.

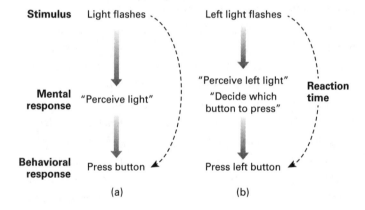

Figure 1.1
Sequence of events between presentation of the stimulus and the behavioral response in Donders's experiments: (a) simple reaction time task; and (b) choice reaction time task. The dashed lines indicate the reaction times Donders measured—the time between the light flash and pressing the button. Note that the mental responses were not measured. Inferences about the mental response were made based on the reaction time measurements.

Think about Donders's method. To determine how long it took to make a decision, he *assumed* that the additional mental activity in the choice task involved deciding which light had flashed and which button to push. He did not actually *observe* people making these decisions, so he *inferred* that these invisible decisions were what caused the slower response.

Donders was joined in the late nineteenth century by other researchers of the mind. The German psychologist Hermann Ebbinghaus (1850–1909) studied memory by determining how accurately lists of nonsense syllables, like IUL, ZRT, or FXP, can be remembered after different delays.[4] The data he collected enabled him to determine "forgetting curves," which plotted the decrease in the number of syllables remembered as a function of the time after they had been presented. Ebbinghaus's results were important because they were one of the first demonstrations that characteristics of a function of the mind (remembering, in this case) could be plotted on a graph.

In the late nineteenth century, the study of the mind was off to a promising start, but just as momentum was building, events occurred that caused the study of the mind to come to a screeching halt. One of the events was Wilhelm Wundt's (1832–1920) founding of the first laboratory of experimental psychology at the University of Leipzig in 1879. Wundt's contributions, treating psychology as a science and establishing psychology as a separate field, were extremely important. But his preferred method, analytic introspection, contributed to the abandonment of the study of the mind in the early twentieth century.

Analytic introspection was a procedure in which a participant was asked to describe his or her experience. For example, in one experiment, Wundt asked participants to describe their experience of hearing a five-tone chord played on the piano. One of the questions Wundt hoped to answer was whether his participants were able to hear each of the individual notes that made up the chord.

Although "self-report" data of this kind were to reappear in psychology laboratories more than one hundred years later, the results from Wundt's lab turned out to be highly variable from participant to participant. This variability bothered John B. Watson (1878–1958), who in 1900 was a graduate student in the psychology department at the University of Chicago. He also did not like that researchers had no way to check the accuracy of the participants' description of their experience. Watson decided, therefore, that some changes were in order if psychology was going to be considered scientific, and he proceeded, with great enthusiasm, to banish the study of the mind from psychology.

The Study of the Mind Is Suspended by Behaviorism

Progress on studying the mind was put on hold in the early 1900s by Watson's founding of *behaviorism*. The flavor of behaviorism is captured in the following quote from Watson's 1913 paper "Psychology as the Behaviorist Views It":

> Psychology as the Behaviorist sees it is a purely objective, experimental branch of natural science. Its theoretical goal is the prediction and control of behavior. *Introspection forms no essential part of its methods,* nor is the scientific value of its data dependent upon the readiness with which they lend themselves to interpretation in terms of consciousness.... What we need to do is start work upon psychology *making behavior, not consciousness, the objective point of our attack.*[5]

In this passage, Watson rejects introspection as a method and proclaims that observable behavior, not events occurring in the mind (which involves unobservable processes such as thinking, emotions, and reasoning), should be psychology's main topic of study. To emphasize his rejection of the mind as the topic of study, he further stated that "psychology...need no longer delude itself into thinking that it is making mental states the object of observation" (163).

In other words, Watson restricted psychology to behavioral data and rejected the idea of going beyond those data to draw conclusions about unobservable mental events. Watson's most famous research paper described his "Little Albert" experiment, in which Watson and Rosalie Rayner[6] used the classical conditioning procedure also used by Ivan Pavlov.[7] In experiments begun in the 1890s, Pavlov paired food and a bell to cause dogs to salivate when they later heard the bell. Watson paired a loud noise with a small rabbit, which Albert had previously liked, to cause Albert to become afraid of the rabbit. Conditioning, according to Watson, provided explanations for many human behaviors, without having to make inferences about what is going on in the mind.

As behaviorism became the dominant force in American psychology, psychologists' attention shifted from asking "What does behavior tell us about the mind?" to "What does the way people or animals react to stimuli tell us about behavior?"

Later, B. F. Skinner (1904–1990) developed a procedure for measuring behavior called *operant conditioning* in which rats and pigeons pressed bars to receive food rewards. Using this procedure, Skinner was able to demonstrate relationships between "schedules of reinforcement"—how often and in what pattern rewards were dispersed when animals pressed the bar—and the animals' bar-pressing behavior. For example, if a rat is rewarded every time it presses the bar, its rate and pattern of bar pressing will be different from when it is rewarded every fifth time it presses the bar.[8]

The beauty of Skinner's system was that it was objective and therefore "scientific." Operant conditioning also led to practical applications for humans such as "behavior therapy," in which the therapist applies Skinner's reward principles to change a patient's unwanted behaviors. However, beginning in the mid-1950s, changes were occurring in both psychology departments and popular culture that began the resurrection of the study of the mind.

A Paradigm Shift toward the Study of the Mind

As behaviorism was dominating psychology in the 1950s, there were stirrings indicating that a paradigm shift was about to occur in psychology, where a paradigm is a system of concepts and experimental procedures that dominate science at a particular time, and a paradigm shift is a change from one paradigm to another.[9]

An example of a paradigm shift in science is the shift from classical physics (associated with the work of Isaac Newton and other eighteenth- and nineteenth-century researchers) to modern physics (associated with Einstein's theory of relativity and the development, by others, of quantum theory) that occurred at the beginning of the twentieth century. The paradigm shift in psychology was the shift from behaviorism, which focused solely on observable behavior, to cognitive psychology, which took the giant step of using observable behavior to make inferences about the operation of the mind.

An event that played an important role in triggering the shift from behaviorism to cognitive psychology occurred in 1954, when IBM introduced a computer that was available to the general public. These computers were still extremely large compared to the laptops of today, but they found their way into university research laboratories, where they were used both to analyze data and, most important for our purposes, to suggest a new way of thinking about the mind.

One of the characteristics of computers that captured the attention of psychologists was that computers processed information in stages, as illustrated in figure 1.2a. In this diagram, information is first received by an "input processor." It is then stored in a "memory unit" before it is processed by an "arithmetic unit," which then creates the computer's output. Using this stage approach as their inspiration, some psychologists proposed the *information-processing approach* to studying the mind. According to the information-processing approach, the operation of the mind can be described as a sequence of mental operations.

The diagram in figure 1.2b is an example of an early flow diagram of the mind, created by the British psychologist Donald Broadbent in 1958.[10] This diagram was inspired by experiments designed to test people's ability to focus on one message, when other messages are presented at the same time, as might occur if you are talking to a friend at a party, while ignoring all the other conversations that are occurring around you. This situation was studied in the laboratory by Colin Cherry,[11] who presented

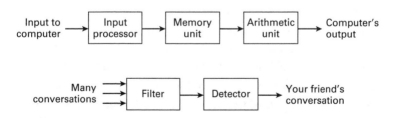

Figure 1.2
Top: Simplified computer flow diagram; bottom: Broadbent's flow diagram of the mind.

participants with two auditory messages, one to the left ear and one to the right ear, and told them to focus their attention on one of the messages (the *attended message*) and to ignore the other one (the *unattended message*). For example, the participant might be told to attend to the left-ear message, which began with "Sam was looking forward to seeing his family over vacation...," while simultaneously receiving, but not attending to, the right-ear message, "The mind is, according to some people, what makes us human..."

When people focused on the attended message, they could clearly hear it, and while they could tell that the unattended message was present, they were unaware of its contents. The flow diagram in figure 1.2b represents Broadbent's idea of how this process might be occurring in the mind. The "input" represents many messages, as might be present at a party. The messages enter the "filter," which lets through what your friend is saying while filtering out all the other conversations. Your friend's message then enters the "detector," and you hear what she is saying.

Broadbent's flow diagram provided a way to analyze the operation of the mind in terms of a sequence of processing stages and proposed a model that could be tested by further experiments. But Cherry and Broadbent were not the only researchers pursuing new ways of studying the mind. At about the same time, John McCarthy, a young professor of mathematics at Dartmouth College, had an idea. Would it be possible, McCarthy wondered, to program computers to mimic the operation of the human mind? To address this question, McCarthy organized a conference at Dartmouth in the summer of 1956 to provide a forum for researchers to discuss ways that computers could be programmed to carry out intelligent behavior. The title of the conference, "Summer Research Project on Artificial Intelligence," was the first use of the term "artificial intelligence." McCarthy defined the artificial intelligence approach as "making a machine behave in ways that would be called intelligent if a human were so behaving."[12]

Researchers from a number of different disciplines—psychologists, mathematicians, computer scientists, linguists, and experts in information theory—attended the conference, which spanned ten weeks. Two participants—Herb Simon and Alan Newell from the Carnegie Institute of Technology—demonstrated a computer program, called the Logic Theorist, at the end of the conference. What they demonstrated was revolutionary, because the Logic Theorist program was able to create proofs of mathematical theorems that involved principles of logic. This program, although primitive compared to modern artificial intelligence programs, created a real "thinking machine" because it did more than simply process numbers; it used humanlike reasoning processes to solve problems.

Shortly after the Dartmouth conference, in September of the same year, another pivotal conference was held, the "Massachusetts Institute of Technology Symposium on Information Theory." This conference provided another opportunity for Newell and Simon to demonstrate their Logic Theorist program, and the attendees also heard George Miller, a Harvard psychologist, present a version of his paper "The Magical Number Seven, Plus or Minus Two," which had just been published.[13] In that paper, Miller presented the idea that our ability to process information has certain limits—that the information processing of the human mind is limited to about seven items (for example, the length of a telephone number not including area code).

The events I have described—Cherry's experiment, Broadbent's filter model, and the two conferences in 1956—represented the beginning of the paradigm shift in psychology that has been called the *cognitive revolution*. It is worth noting, however, that the shift from behaviorism to the cognitive approach, which was indeed revolutionary, occurred over a period of time. The scientists attending the conferences in 1956 had no idea that these conferences would, years later, be seen as historic events in the birth of a new way of thinking about the mind or that scientific historians would someday call 1956 "the birthday of cognitive science."[14]

Ironically, another development that opened the mind as a topic of study was the publication, in 1957, of a book by B. F. Skinner titled *Verbal Behavior*.[15] In this book, Skinner argued that children learn language through operant conditioning. According to this idea, children imitate speech that they hear, and repeat correct speech because it is rewarded. But in 1959, Noam Chomsky, a linguist at MIT, published a scathing review of Skinner's book, in which Chomsky pointed out that children say many sentences that have never been rewarded by parents ("I hate you, Mommy," for example), and that during the normal course of language development, they go through a stage in which they use incorrect grammar, such as "the boy hitted the ball," even though this incorrect grammar may never have been reinforced.[16]

Chomsky saw language development as being determined not by imitation or reinforcement but by an inborn biological program that holds across cultures. Chomsky's idea that language is a product of the way the mind is constructed, rather than a result of reinforcement, led psychologists to reconsider the idea that language and other complex behaviors, such as problem-solving and reasoning, can be explained by operant conditioning. Instead they began to realize that to understand complex cognitive behaviors, it is necessary not only to measure observable behavior but also to consider what this behavior tells us about how the mind works.

As more psychologists became interested in studying the mind, Ulrich Neisser published the first textbook titled *Cognitive Psychology* in 1967,[17] and psychologists studying the mind began calling themselves "cognitive psychologists." More flow diagrams followed, describing processes ranging from memory to language to problem-solving in terms of information processing, and as researchers embraced the information-processing approach, with its commitment to discovering the internal mechanisms of the mind, behaviorism began fading into the background.[18]

Meanwhile in Popular Culture...

The cognitive revolution was, for our purposes, the major psychological event of the 1950s and 1960s, because it paved the way for the reentry of the study of the mind into psychology. But during that period, another revolution was taking place: the *sixties counterculture revolution*. On the one hand, this revolution, in which some of the participants were hippies getting high during the Summer of Love in San Francisco, would seem unrelated to research being carried out in psychology laboratories. However, as chronicled in Adam Smith's book *Powers of Mind*, the sixties revolution and the cognitive revolution had an important thing in common: they were both concerned with the mind.[19]

To appreciate why events that were occurring in society in the 1960s were called a revolution, let's consider what was happening in society in the previous decade. In 1953, Dwight D. Eisenhower was president of the United States, and families were pursuing the American Dream of home ownership in newly developing suburban communities, which were accompanied by the development of shopping centers and fast-food restaurants, like McDonalds, which opened its first restaurant in 1956. Emblematic of the 1950s was the TV sitcom *Leave It to Beaver*, which premiered in 1957 and ran for six seasons. This sitcom featured the Cleaver family, which was the wholesome "ideal family" of the 1950s and 1960s TV, where dad Ward Cleaver arrives home from work and is greeted by mom June Cleaver, who has spent her day cleaning, cooking, and tending to the needs of their two sons.

But the idealized Cleaver family was not to last as a symbol of American normalcy. The United States entered the Vietnam War in the early 1960s, leading to widespread protests. Timothy Leary, then a professor at Harvard, began experimenting with the hallucinogenic drug lysergic acid diethylamide (LSD) and preaching his credo "Turn on, tune in, and drop out," and marijuana use was becoming widespread among young people, both those who identified as part of the hippie counterculture and those who were simply members of the younger generation—who, four hundred thousand strong, showed up at the drug-drenched Woodstock music festival in the summer of 1969. The reason these events are relevant to the study of the mind is that one of the themes of the new counterculture of the 1960s was "mind expansion." Drugs were one way of expanding the mind, and although many used drugs just for "fun," there were also those who saw drugs as an entryway to discovering higher processes of the mind.[20] Additionally, while hippies were getting high, scientists began studying LSD to determine how it works and how it could be used therapeutically.[21] This research halted when psychedelics were made illegal in the United States in the late 1960s. But recently, researchers have begun applying twenty-first-century techniques to determine connections between drug experiences and physiological events in the brain.[22]

Another popular movement associated with the mind was the "human potential movement," headquartered at the Esalen Institute. Founded in 1962 on a campus in Big Sur, California, perched high above the Pacific Ocean 160 miles south of San Francisco, Esalen featured self-awareness seminars, lectures, and activities such as yoga and meditation. Esalen, which still exists today, spawned many similar institutions across the country.

Particularly important among the activities nurtured at Esalen was meditation. Although many Americans in the sixties considered meditation "exotic" or "far out," today it is widely practiced, and its mechanisms and benefits have been studied in thousands of research studies.[23]

These stories about the sixties—the cognitive revolution, which conceptualized the mind as an information processor; the popularization of psychoactive drugs; and the human potential movement, which included meditation practice—each involved the mind in some way. They also have something else in common: they involve phenomena that are created by the brain. "But, of course," you might say, "isn't everything we experience created by the brain?" My answer to this question, as you might expect, is "yes." But before we begin considering research on the brain, let's consider the idea, proposed by some, that the answer is "no."

Mind-Brain Skepticism

The philosophical standard-bearer for the idea of a separation between mind and brain is René Descartes (1596–1650). Descartes's stance relevant to the mind and the brain is called *Cartesian dualism*. He stated that the mind and the brain are made up of different "substances." He located the pineal gland, at the base of the brain, as the place where the mind and brain interact, but he provided no details as to how this interaction might occur. The following reasoning behind the idea that the mind and the brain are two distinct substances appears in his *Discourse on the Method*:[24]

1. I can pretend that I have no body or that there is no world out there.
2. But because I can think, I can't pretend that I don't exist.

From this second idea comes the famous pronouncement *Cogito ergo sum*, "I think, therefore I am." Descartes goes on to say that he is "a substance whose essence or nature is simply to think, and which does not require any place, or depend on any material thing, in order to exist" (6:32–33). The key thought here is that thinking does not depend on a "material thing," and because the brain is a material thing, Descartes concluded that the thinking mind does not depend on the brain.

God and spirituality played a central role in Descartes's philosophy, so it should not surprise us that he points to God as the source of both mind and body. Other spiritually centered positions have also subscribed to the separateness of mind and brain. For example, Geshe Kelsang Gyatso,[25] a contemporary Buddhist monk, states: "Our brain is not our mind. The brain is simply a part of our body that, for example, can be photographed, whereas our mind cannot." Another spiritualist approach that denies the necessity of the brain is taken by Deepak Chopra, a follower of Vedantic Hindu philosophy and a widely read author.[26] He does not deny some role for the brain but emphasizes that because biological research has left many questions about the mind unanswered, we need to look beyond the brain to understand the mind, to properties of the "conscious universe." Chopra's reasoning, which is not easy to follow, leads him to conclude that everything in the universe is conscious—not just organisms with brains. (In chapter 2, we will see that consciousness is classified as a product of the mind, so when Chopra talks about consciousness, he is essentially referring to the mind.)

Another way the mind and brain have been seen as separate derives from descriptions of *out-of-body experiences* (OBEs). Out-of-body experiences occur when a person seems to be awake and feels that his or her self, or center of experience, is located outside of his or her body. Sometimes also occurring during an OBE is *autoscopy*, in which the person experiences his or her body as floating in space.[27] Here is an example of one person's OBE:

> I was in bed and about to fall asleep when I had the distinct impression that "I" was at the ceiling level looking down at my body in the bed. I was very startled and frightened; immediately afterwards I felt that I was consciously back in the bed again.[28]

OBEs have been associated with psychiatric conditions such as schizophrenia, depression, and personality disorder, as well as with neurological problems such as epilepsy, and can also occur in 10 percent of the general public.[29]

There is no controversy regarding the existence of OBE, but there is controversy regarding its cause. There are two opposing points of view. One view states that OBE represents the projection of a person's personality into space and is therefore an example of a separation of mind from body. This spiritual explanation of OBE is not, however, accepted by most psychologists, who explain OBE as the result of normal psychological and physiological processes.

Evidence connecting OBEs to brain processes includes the finding that drugs like marijuana and LSD increase the probability of experiencing an OBE, presumably through the drug's action on the brain. Dirk De Ridder and coworkers demonstrated the brain-OBE link in a patient who was undergoing treatment for tinnitus (ringing in the ears) that included electrical stimulation of a location in the temporal lobe.[30] Unfortunately, this stimulation did not relieve the patient's tinnitus symptoms. However, it did cause the person to experience an OBE, and when the person's brain was scanned during the OBE, areas in the temporal lobe were activated that are involved in the somatosensory system (sensing the body surface) and in the vestibular system (regulating the sense of balance). Findings such as these have led to the hypothesis that OBEs are caused by distorted somatosensory and vestibular processing.[31]

Sometimes OBE occurs in conjunction with the phenomenon of near-death experience (NDE). This phenomenon has been reported by people, such as cardiac arrest patients, who were near death and then were revived. They typically report experiencing OBE and seeing a tunnel, visions of a brilliant white or gold light, meeting other beings, and seeing their life passing by. The spiritualist explanation for NDE is that the person has left his or her body and experienced a higher spiritual world. This is sometimes called the "afterlife hypothesis."[32]

Eben Alexander, a neurosurgeon, had an NDE and wrote a book, *Proof of Heaven*, about his experience and why it occurred.[33] His explanation hinges on his assertion that "during my coma, my brain wasn't working improperly—it wasn't working at all." In other words, according to Alexander, his NDE occurred when his brain was dead. From this he concluded that NDEs show that experience does not depend on the brain, and at least in his case, his experience occurred because he was actually in heaven. Others, such as Pim van Lommel in his book *Consciousness beyond Life*, have used similar reasoning to argue that the brain is not necessary for experience.[34] Ironically, the subtitle of Van Lommel's book is *The Science of Near-Death Experience*. And here's the rub—although both Alexander and Van Lommel are physicians, the "science" they cite is simply inaccurate or speculative.[35] A basic question that strikes at the central argument in both books is "Was the brain 'dead' when people were having NDEs?" After all, these people did live to tell their story, which means their brain was alive, as they were "coming back." Serious scientific research on NDE indicates that patients' experiences are most likely to occur as they are slowly regaining consciousness, and so the supposed "spiritual" experiences are due to brain processes occurring as the brain is becoming functional again.[36]

One way to think about NDEs and OBEs is that they are a type of hallucination created by the brain. The neurologist Oliver Sacks, in writing about Alexander's book, rejects the idea that his vision was "nonphysiological." In considering hallucinations in general, Sacks says:

> Hallucinations, whether revelatory or banal, are not of supernatural origin; they are part of the normal range of human consciousness and experience. This is not to say that they cannot play a part in the spiritual life, or have great meaning for the individual. Yet while it is understandable that one might attribute value, ground beliefs, or construct narratives from them, hallucinations cannot provide evidence for the existence of any metaphysical beings or places. They provide evidence only of the brain's power to create them.[37]

The idea that hallucinations are created by the brain is the centerpiece of research I mentioned earlier, which has begun to search for links between physiological changes in the brain caused by drugs like LSD and the drug-induced experiences, which are often described as "spiritual."[38]

We have seen that, beginning with Descartes and continuing to the present, some people feel that experience does not depend solely on the brain. But most cognitive neuroscientists reject this idea in favor of the idea that all experience is created by the brain.[39] The theme of this book is that although we still have much to learn about the mind, traditional scientific approaches offer the best pathway for figuring it out. It is relevant that in John Brockman's 2013 book *The Mind*, which included articles by eighteen leading scientists, sixteen of the eighteen articles specifically mentioned the brain, and the other two described empirically based behavioral research.[40] Brockman's book reflects the preeminence of the scientific, empirical, and often biologically oriented approach to the study of the mind. It also reflects the approach of this book. We will focus on empirical research—with a little speculation thrown in—and will be looking at links between the brain and mind whenever possible.

Mind-Brain Connections

> The human mind is a complex phenomenon built on the physical scaffolding of the brain.
> —Danielle Bassett and Michael Gazzaniga[41]

How is our invisible mind created from the physical scaffolding of the brain? Answers to this question began appearing in the 1800s, when the predominant technique for determining mind-brain connections was analyzing the behavior of patients with brain damage. In 1861, Paul Broca (1824–1880) reported his study of a patient who suffered damage to his frontal lobe and was called "Tan" because that was the only word he could say.

When Broca tested other patents with damage to their frontal lobe, in the area that came to be called Broca's area,[42] he found that their speech was slow and labored and often had jumbled sentence structure. Modern researchers have concluded that damage to Broca's area causes problems creating meaning based on word order. Another area relevant to speech was identified by Carl Wernicke (1848–1905), who studied patients with damage to an area in their temporal lobe now called Wernicke's area.[43] Patients with damage in this area have difficulty understanding the meanings of words. The classic studies of Broca and Wernicke were precursors to modern research in neuropsychology: the study of the behavior of people with brain damage. Figure 1.3a shows the locations of Broca's and Wernicke's areas.

The next major advance, looking farther "under the hood" of the brain, opened the way for work at the level of the neuron. The Spanish physiologist Santiago Ramón y Cajal (1852–1934), peering through his microscope at exquisitely stained slices of brain tissue, discovered that individual units called neurons were the basic building blocks of the brain (fig. 1.3b). Cajal also concluded that neurons communicate with one another to form neural circuits.[44]

Cajal's idea of individual neurons that communicate with other neurons to form neural circuits laid the groundwork for later research on neural communication, which I discuss in more detail in chapter 6. These discoveries earned Cajal the Nobel Prize in 1906, and today he is recognized as "the person who made this cellular study of mental life possible."[45]

Cajal succeeded in describing the structure of individual neurons and how they are related to other neurons, and he knew that these neurons transmitted electrical signals. However, although Cajal was able to *see* individual neurons, it wasn't yet possible, early in the twentieth century, to measure the electrical signals that traveled down these neurons. Scientists faced two problems: (1) neurons are very small, and (2) so are the electrical signals that travel down the neurons.

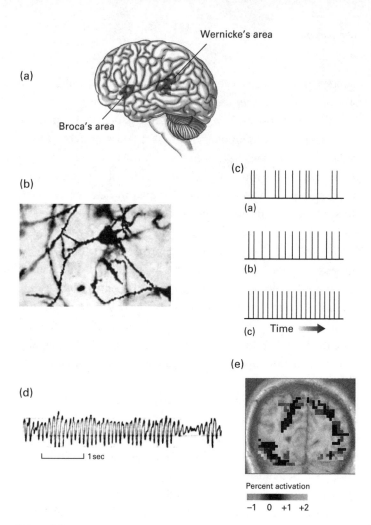

Figure 1.3

Results of some physiological research on the mind. (a) Location of Broca's and Wernicke's areas; (b) A neuron and some nerve fibers as visualized by Cajal. (c) Modern records of nerve impulses from a single neuron. The rate of nerve firing increases as stimulus intensity increases, from top to bottom. (d) An electroencephalogram (EEG) recorded with scalp electrodes. (e) fMRI record. Activity is determined for each voxel, where a voxel is a small volume of the cortex, indicated here by small squares. Colors, not shown here, indicate the amount of activity in each voxel.

In 1906, when Cajal received his Nobel Prize, the technology for measuring small responses in small neurons was not available, but by the 1920s techniques had been developed to isolate single neurons, and a device called the "three-stage amplifier" became available. With the aid of this new technology, the British physiologist Edgar Adrian (1889–1977) founded modern single-neuron electrophysiology—the recording of electrical signals from single neurons—an achievement for which he was awarded the Nobel Prize in 1932.[46]

Once Adrian had succeeded in recording electrical signals from neurons, the stage was set for research linking brain activity and experience. In an early experiment, Adrian recorded from a neuron receiving impulses from the skin of a frog. As he applied pressure to the skin, he found that pressure caused rapid signals called nerve impulses, and increasing the pressure caused the *rate* of nerve firing—the number of nerve impulses that traveled down the nerve fiber per second—to increase (fig. 1.3c). From this result, Adrian drew a connection between nerve firing and experience. He described this connection in his 1928 book *The Basis of Sensation* by stating that if nerve impulses "are crowded closely together the sensation is intense, if they are separated by long intervals the sensation is correspondingly feeble."[47] What Adrian is saying is that electrical signals are *representing* the intensity of the stimulus, so pressure that generates "crowded" electrical signals feels stronger than pressure that generates signals separated by long intervals.

Determining that faster nerve firing signaled more pressure was the first step toward determining the relation between electrical signals in the brain and experience and led to tens of thousands of papers using the single-neuron recording technique.[48]

Most of the single-neuron research was done on animals. The discovery of the electroencephalogram by Hans Berger in

1929 made it possible to record electrical signals by placing disc electrodes on the scalps of humans.[49] These electrodes recorded the massed response of many neurons and led to research relating brain activity in humans to various states of consciousness, which I discuss in chapter 2 (fig. 1.3d).

Another technological advance that has furthered our understanding of mind-brain connections is the development of techniques for brain imaging. The first imaging technique, *positron emission tomography (PET)*, was introduced in 1975, and was largely replaced by *functional magnetic resonance imaging (fMRI)* which was introduced in 1990.[50] Functional magnetic resonance imaging is based on the fact that blood flow increases in areas of the brain activated by a cognitive task. Without going into the details, these changes in blood flow, measured in a brain scanner, are converted into images that indicate which areas of the brain become more activated by a particular task, and which areas become less activated (fig. 1.3e).

The introduction of brain imaging brings us back to the idea of paradigm shifts. The idea of paradigm shifts, introduced by Thomas Kuhn in his 1962 book *The Structure of Scientific Revolutions*, was based on the idea that a scientific revolution involves a shift in the way people think about a subject.[51] This was clearly the case in the shift from the behavioral to the cognitive paradigm. But in addition to the shift in thinking, another kind of shift can occur: a shift in how people *do* science.[52] This shift, which depends on new developments in technology, is what happened with the introduction of fMRI. *NeuroImage*, a journal devoted solely to reporting neuroimaging research, was founded in 1992, followed by *Human Brain Mapping* in 1993.[53] From its starting point in the early 1990s, the number of fMRI papers published in all journals has steadily increased. It has been estimated

Table 1.1
Physiological Methods for Studying Mind and Brain

Method	Early Work
(a) Neuropsychology Study of the effect of brain damage on human behavior	**1861:** Broca; **1868:** Wernicke. Specific language functions are located in specific brain areas. See fig. 1.3a.
(b) Neuroanatomy Determination of structures in the nervous system and their connections	**1894:** Ramón y Cajal. Neurons create the transmission system of the brain. Often organized in neural circuits. See fig. 1.3b.
(c) Single-cell electrophysiology Recording electrical signals from single neurons with microelectrodes and determining how they fire to sensory stimuli (mainly in animals)	**1928:** Edgar Adrian. Demonstrated a relation between rate of nerve firing and magnitude of sensory stimuli. See fig. 1.3c.
(d) Electroencephalography Recording electrical signals from the surface of the human scalp.	**1929:** Hans Berger. Brain wave patterns are related to states of consciousness, especially during sleep. See fig. 1.3d.
(e) Brain imaging Measuring change in blood flow in human brain caused by cognitive activity.	**1975:** Ter-Pogossian et al. Positron-emission tomography (PET) **1990:** Ogawa et al. Functional magnetic resonance imaging (fMRI): Specific areas of the brain are associated with specific functions. Also demonstrated widespread brain activation associated with even simple functions. See fig. 1.3e.

that about forty thousand fMRI papers had been published as of 2015.[54] As we will see in later chapters, fMRI has played a central role in studying the connection between brain function and experience.

Table 1.1 summarizes some of the physiological methods used to study mind-brain connections. As we will see in later chapters, other imaging techniques have been developed, which provide additional information about brain structure and function. In the next chapter, I focus on consciousness—the subjective inner life of the mind. The focus will be on experience and behavior, but we will return to the brain at the end of the chapter.

2 Consciousness and Experience

As you read this, it is likely you are conscious. After all, you are seeing words and are aware of their meanings. You may also be aware of your book or computer screen and some of the things happening around you. Meanwhile, as you are conscious of some things, you are not conscious of others. Until right now, you probably weren't paying attention to the feeling of your clothes on your body, and you certainly aren't directly aware of the firing of neurons in your brain.

What Is Consciousness?

> The problem of consciousness is arguably the central issue in current theorizing about the mind.
> —Robert Van Gulick[1]

This chapter moves beyond introducing the mind to introducing one of the mind's most impressive and baffling creations: consciousness. In fact, the definition of *consciousness* is so elusive that there is no generally accepted definition.[2] However, here are a few proposals. Consciousness has been defined as:

the perception of what passes in a man's own mind[3]
[the] subjective inner life of the mind[4]
what experience feels like from the inside[5]
your private experience[6]
the ability to be aware of being aware[7]
all of our states of feeling or sentience or awareness[8]

The definition we will adopt is a commonsense definition accepted by many researchers. We define consciousness as *what your own personal experience feels like from the inside*; or, put another way, consciousness is the *subjective inner life of the mind*. Thus, returning to the example that opened the chapter, how it *felt* to be reading the words in your book or on your computer screen would fit this "personal" definition of consciousness, which, because it refers to what *you* are experiencing, is called the *first-person approach to consciousness*.

The experiences or feelings associated with the first-person approach are called *qualia* (singular: *quale*), where qualia are the raw essence of an experience. Thus, when you look at a red tomato, the redness you experience is a red quale. Someone else looking at the same tomato would experience his or her own red quale. Are your qualia and the other person's the same? We will return to that question later in the chapter.

Although looking at a tomato may take place at one point in time, consciousness has also been described as continuing through time. One way this dynamic aspect of consciousness has been described is by comparing it to a movie. "There is," the philosopher David Chalmers says, "an amazing movie that seems to be playing out in our minds." This movie consists of images, sounds, thoughts, and emotions.[9] Consciousness has also been described as playing out on a stage—with particularly vivid experiences highlighted by a spotlight.[10]

Another dynamic conception of consciousness, proposed by William James in 1890, compares consciousness to a river or stream:

> Consciousness, then, does not appear to itself chopped up in bits. A "river" or "stream" are the metaphors by which it is most abundantly diluted. In talking of it hereafter, let us call it the stream of thought, of consciousness, or of subjective life.[11]

A movie, a play on a stage, a floating stream, or just "what I'm experiencing right now": however consciousness is described, it is about having and being aware of experiences. So if you are awake and aware or if you are asleep and dreaming, you are conscious. And it's easy enough to tell, because you are extremely familiar with your own experiences. After all, what could be more familiar than all the things you are experiencing in your day-to-day life? It is this familiarity that Anil Seth is referring to when he states: "Consciousness is at once the most familiar and the most mysterious aspect of existence."[12]

But if consciousness is so familiar, why does Seth also say that it is mysterious? He goes on to say: "Conscious experiences define our lives, but the subjective, private and qualitative nature of these experiences seems to resist scientific inquiry."

Consciousness, then, is familiar, but because it is so difficult to study, it is also mysterious. What could be more interesting than something that is both familiar *and* mysterious? In this chapter, you will become aware of these mysteries that make consciousness difficult to study as we discuss the following four questions:

1. Are nonhuman animals conscious?
2. What is conscious experience like in nonhuman animals?

3. What is conscious experience like in another person?
4. How does the nervous system create our experiences?

We begin with the first question.

Consciousness in Nonhumans

One of my favorite days in my undergraduate cognitive psychology class is when we consider the possibility of nonhuman consciousness. On that day, students bring in stuffed animals (it's always amazing what some students have hidden away in their dorm rooms), which we use to create a "consciousness lineup."

The lineup begins with a human (me), who from the student's viewpoint is at the far left end. Then, extending to the right, the task is to line up the animals *in order of consciousness*. So, from the class's point of view, it's like reading—beginning with the most conscious on the left, the animals become less and less conscious, moving along the lineup to the right, with the class's task being to determine the correct order. Chimpanzees and monkeys come first, followed by dogs, cats, and other animals. Oh, yes, I also bring a plant and a rock to class, which students place at the far right (least conscious) end of the lineup.

The class is faced with two questions about the lineup: (1) is there a dividing line between conscious and nonconscious, and if so, where is it; and (2) what is consciousness like for each of the animals along the line? Let's begin with question 1: determining the dividing line. Two extreme positions about the dividing line have been proposed. Descartes's statement "I think, therefore I am," staked out consciousness as the sole province of humans, because we are the thinking animal. Other animals are, according to Descartes, machinelike automatons. The line between conscious and not conscious would, for Descartes, be right next to humans. Few people today would agree with this idea, as evidence suggests that many animals do have consciousness.

At the other extreme is *panpsychism*, the idea that consciousness is a property of the universe, so everything is conscious, including not only humans and all animals but also plants, rocks, and everything else. Some people take this idea seriously;[13] however, belief in panpsychism demands faith, as no scientific evidence supports it.[14]

With the correct answer being somewhere between Descartes and panpsychism, we can state that a line divides conscious and nonconscious somewhere between humans and rocks. Let's agree that rocks are not conscious. But what about plants?

Are Plants Conscious?

Asking whether plants are conscious may sound like a silly question, when we remember that the main requirement for having consciousness is having an inner life, which includes an awareness of "experience." But Peter Tompkins and Christopher Bird, the authors of *The Secret Life of Plants*, did not consider that question silly in 1973.[15] They put forth the idea that plants preferred classical music to rock and roll, and that they could think, feel emotions, and respond to what humans were thinking. This book was published at the right time—when New Age thinking was becoming part of the culture—and landed on the *New York Times* best seller list for nonfiction.

As it turned out, legitimate plant scientists discredited just about all the "evidence" cited in the book (which, as it turns out, belonged on the fiction list). But the harm had been done to the field of plant science. As Michael Pollan noted, "Americans began talking to their plants and playing Mozart for them, and no doubt many still do."[16]

After this setback to legitimate plant science, many researchers hesitated to draw parallels between plants and humans. However, legitimate research on this topic eventually began to appear. The renowned plant biologist Daniel Chamovitz summarizes much current research on plants in his book *What a Plant Knows*. In this book, he describes the sensory and "cognitive" capacities of plants in chapters like "What a Plant Sees," "What a Plant Feels," and "What a Plant Remembers." It is a clever book because it draws parallels between humans and plants, both in terms of shared DNA and in terms of the similarities between human and plant senses.[17]

The science in Chamovitz's book is accurate, and his descriptions of how plants are affected by what is happening around them are engagingly written. But Chamovitz's decision to "humanize" plants by referring to what plants "see," "feel," and "remember" creates the impression that plants are more like humans than they actually are. In fairness to Chamovitz, he notes that when he says "what a plant knows," "what a plant sees," or "what a plant feels," his use of the words *knows*, *sees*, and *feels* is unorthodox, because a plant doesn't have a brain and so doesn't *know*, *see*, or *feel* the way humans do. Nonetheless, words are important, and saying that a plant *sees* implies that plants are conscious of light. The reality, however, is that plants *sense* light, but we have no evidence that they are conscious of it; some plants, like the Venus flytrap, can *react* to the pressure of a fly that makes the mistake of alighting inside it, but the plant doesn't *feel* anything. So with apologies to all those behaving but nonconscious plants out there, we conclude that plants are on the nonconscious side of the dividing line between conscious and nonconscious.[18]

Consciousness in Nonhuman Animals

Meanwhile, back in class, students are arranging animals in order of consciousness and are trying to figure out where the conscious-nonconscious line might be. Most are quick to assign consciousness to chimpanzees, monkeys, dogs, and cats but then start wondering which other animals to include. On what basis can they decide? This question has been approached in two ways: behaviorally and physiologically.

By observing animals, we can identify many behaviors that seem to be associated with positive and negative emotional feelings. Most dog owners have experienced that big-eyed "please forgive me" look or the subservient body language that follows bad behavior (such as eating food from the dinner table). Emotional behaviors, also called affective behaviors, can also be expressed vocally, as in a dog's whining or whimpering.

Affective behaviors are easy to interpret as showing that dogs and other animals are experiencing feelings and are therefore conscious. However, as obvious as their feelings may appear, we have no way of knowing what they are actually feeling, and we need to be wary of engaging in *anthropomorphism*—assigning human characteristics to animals. Observing animals' affective behaviors therefore provides only suggestive evidence of an animal's consciousness.[19]

Another type of observation involves intelligent behavior. There are countless examples of intelligent behaviors in animals. Research on animal learning has demonstrated animals' abilities to learn, communicate, and solve problems. Honeybees engage in "waggle dances" to alert other bees to the location of a food source.[20] Birds exhibit memory, the ability to learn about their social environment and solve problems,[21] and the octopus—an animal most people don't think of when considering intelligent behavior—is perhaps the most intelligent invertebrate.

Octopuses have been in the news because of their ability to escape from zoo aquariums. One escape artist, named Inky, who lived at the National Aquarium of New Zealand, slipped through a small gap at the top of his tank, slithered eight feet across the floor, and slid down a 164-foot-long drainpipe that dropped him into the ocean. The aquarium's manager, commenting on the escape, said, "Didn't even leave us a message."[22]

But octopuses are not merely escape artists. They can navigate complex mazes, determine how to open "puzzle jars" containing tasty crabs, and engage in play activities like blowing jets of water to propel a floating object toward another jet that returns it—"the aquatic equivalent of bouncing a ball."[23] Additionally, recent research has shown that the octopus has an unusually large genome, rivaling humans in the number of genes,[24] which has led to the suggestion that the complexity of the octopus's genome is the basis of its highly developed cognitive skills.[25]

Do behaviors like these imply some form of consciousness? Again, we have to beware of anthropomorphism, and there is certainly no guarantee that animals that behave like humans have human-like experiences.[26] Thus, in addition to using behavioral observations to determine the presence of consciousness, researchers also consider physiological functioning.

One argument for animals having consciousness is based on a comparison of animal and human brains. The human brain is distinguished by a large cerebral cortex (fig. 2.1), which overlays the brain's more primitive subcortical structures, including the midbrain. As we move down the phylogenetic scale, animals' cerebral cortices become smaller, and eventually subcortical structures dominate. Many animals lower on the phylogenetic scale, such as birds and insects, have structures that serve functions similar to the human midbrain.[27]

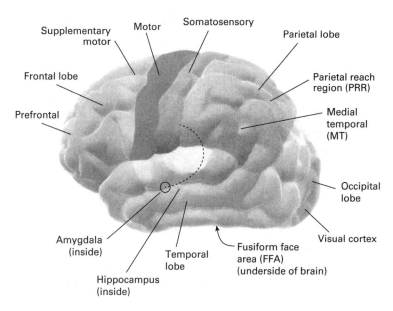

Figure 2.1
Human brain showing locations of structures and areas that are referred to in this book.

Human conscious experience is usually associated with higher-level cerebral structures, especially for higher levels of functioning such as abstract reasoning, creating and responding to music, and pondering existential questions like the meaning of life. Thus, if we consider only the cerebral cortex, we might be tempted to conclude that animals that lack a cerebral cortex may also lack consciousness. However, it turns out that some consciousness remains in humans even when the cerebral cortex has been extensively damaged, because the midbrain structures that remain still support subjective experiences like emotions.[28]

Because humans with just a midbrain can have subjective experiences, some scientists have argued that animals like birds and insects, which have structures that function like the human midbrain, can also have subjective experiences.[29] Researchers have also identified neural circuits in birds[30] and honeybees[31] associated with emotional feelings and have shown that neurochemicals associated with emotions in humans are also present in animals.[32]

Although no single behavioral or physiological observation guarantees that an animal is conscious, all the evidence taken together persuaded a group of prominent neuroscientists to release "The Cambridge Declaration on Consciousness" on July 7, 2012, which presents information supporting animal consciousness and ends with the following statement:

> We declare the following: The absence of a neocortex does not appear to preclude an organism from experiencing affective states. Convergent evidence indicates that non-human animals have the neuroemotional, neurochemical and neurophysiological substrates of conscious states.... Consequently, the weight of evidence indicates that humans are not unique in possessing neurological substrates that generate consciousness.[33]

What this means, in brief, is that some forms of consciousness are likely to be present in many species. But the Cambridge Declaration sidesteps the problem of which animals are conscious and which are not. Researchers are still debating exactly where to draw the line indicating which species are conscious.

What Do Animals Experience?

Having concluded that there is a dividing line between conscious and nonconscious, we can move on to the next question: what is consciousness like for animals on the conscious side of the line? Although many different animals may have consciousness, there are undoubtedly vast differences in the nature of conscious experience in different animals. One theory regarding these differences is that the nature of consciousness in a particular animal may be related to the complexity of its neural processing. As stated by David Chalmers: "Where you find complex information processing, you find complex consciousness. As information processing gets simpler, you find some kind of simpler consciousness."[34]

Complex consciousness would include human experience, in which we experience a wide variety of qualia, and in addition can have "higher-order" thoughts on things like the nature of what we are experiencing, or perhaps even the nature of life in general. Simple consciousness might include just a few simple qualia, with no capacity for reflection. Along these lines, a newly developing research literature on the relationship between neural complexity and consciousness is considering this idea in more detail.[35]

A philosophical approach to what consciousness is like for nonhuman animals was put forth by Thomas Nagel in his classic paper titled "What Is It Like to Be a Bat?"[36] Nagel set forth his basic premise as follows: "The fact that an organism has conscious experience *at all* means, basically, that there is something it is like to *be* that organism." Nagel picked bats for his question rather than monkeys, dogs, or cats because of the large differences between the sensory apparatus of bats and humans. The most striking difference is in the way the bat "sees." Bats sense where things are in their environment by sending out high-frequency sonar pulses, which hit objects and bounce back. By sensing how long it takes to receive the reflection, the bat determines the location and distance of the object. This is the process of *echolocation*.

So what is the bat experiencing? When humans look at a scene that contains objects like houses, trees, and cars, we see the scene because light is reflected from each object and into our eyes, and we visually apprehend each object in space and the objects' relations to one another. But when a bat flies through the scene, the main information it receives is not reflected light but rather the reflections of sonar pulses that it produces itself.

Because the bat's pulses are more like the vibrations in the air that humans perceive as sounds (although of a much higher frequency than we can hear), should we assume that the bat "hears" sounds that it transforms into information about the objects in a scene? It doesn't take much thinking about what the bat may or may not be experiencing before it becomes obvious that we have absolutely no way of answering this question. The bat, of course, knows the answer. The bat is experiencing what it is like to be a bat!

It is a little easier to guess what monkeys, dogs, or cats might be experiencing because they are much more like us. But even in these cases, we can only guess what they are experiencing.

All this conjecture about animals brings us to a related question: how can we know what another *person* is experiencing? On the surface, it would seem that answering this question would be much easier than answering the question for other animals, because we can talk to each other and compare notes. But, as we will see next, we really have no way of knowing what someone else is experiencing.

The Mysteries of Human Experience

We are now ready to consider the mystery of human experience, which has the advantage, compared to animals, of being something we can talk to each other about. If you don't consider experience as being mysterious, imagine the following: You are standing at the edge of the Grand Canyon, looking down at the river far below and across the chasm to the other side. Your friend is standing by your side, and you are comparing notes about the awesome scene before you. It is easy to talk about it because you are both looking at the same scene, and from your conversation, it seems as if you are, in fact, seeing the same things.

But suddenly something magical happens: the *experience fairy* taps both of you on the shoulder and causes you to exchange experiences. You are now experiencing the scene as your friend had just experienced it moments before. How does your new experience (previously your friend's) compare to your original experience? This point of this "thought experiment" is that there is no way to answer the question. Although your experience could remain the same after the tap, it is also possible that your experience would change. The change could be subtle ("Hmmm, the colors shifted slightly") or dramatic ("Wow, the colors are different, and the far rim seems much farther away"). It is difficult to know what will happen because everyone's experience is private and is known only to the person who is having the experience.

Can We Know What Someone Else Is Experiencing?

That I am conscious here and now, is the one fact I am absolutely certain of. All the rest is conjecture.
—Giulio Tononi and Christof Koch[37]

Let's further explore the idea that because experience is private, someone else's experience is open to conjecture. We take as our starting point the idea of qualia—the raw essence of an experience, introduced at the beginning of the chapter. Because qualia refers to a person's experience, we can say that everyone's qualia are private. So when you say "I see red," you know exactly what it means because you are experiencing it. But when Susan says "I see red," you can only guess at what she is experiencing because she is talking about her own private qualia. The natural starting point for guessing what Susan's qualia might be like would probably be your own experience, and then you can use clues from other things Susan says, like "that's as bright as Rudolph's glowing nose" or "it's close to the color of blood." But no matter what information you have, you can never be certain that you really know what Susan's red is—or, to put it another way, you can never be certain as to the essence of Susan's qualia.

In my class, I deal with the question of whether we can know other people's qualia through an exercise involving zombies. As we know from novels and movies, zombies are gross looking— they stare into space, walk funny, and often drool. And, oh, yes, they bite people, drink their blood, and turn them into zombies. Finally, zombies are, by definition, dead inside. That is, they have no inner experience and so don't experience qualia. All of this is enough to make us want to stay away from zombies. But the philosopher David Chalmers introduced a different kind of zombie, called *philosophical zombies*, which are more pleasant to be around and offer us the opportunity to think further about the private nature of experience.[38]

Philosophical zombies are different from gross movie zombies because they are exactly like us in appearance and behavior. Thus, if you were to meet a philosophical zombie, you would conclude, based on observing its appearance and behavior, that it was one of "us." But let's not forget that a philosophical zombie is, after all, a zombie, and that all zombies, whether gross movie zombies or presentable philosophical zombies, are dead inside. They have no inner experience and do not experience qualia. Thus they may be able to detect a rose and determine its physical characteristics, but they have no inner experience connected to the rose.

After describing philosophical zombies to my class, I propose the following zombie exercise: "It has come to my attention that there are a few philosophical zombies lurking among us. Your assignment is to pair up with another person and determine whether he or she is a philosophical zombie." Some students throw themselves into this, and animated discussions break out throughout the classroom. Afterward, many students report that their partner is *not* a philosophical zombie because they have reported that they experience emotions ("I was sad when my grandmother died") or sensory experiences ("I love the way a sunset looks").

The students who claim they can tell that someone is not a philosophical zombie are making the error many of us make all the time: they are assuming that when other people say they are experiencing "red" or "sadness" or how they feel about seeing a sunset, they are having experiences similar to our own. Remember, however, that philosophical zombies are just "talking the talk" by acting like us but are still dead inside and thus cannot be experiencing anything, no matter what they say. So I point out to students that the person they thought was not a philosophical zombie could, in fact, be one.

The philosophical zombie exercise goes a long way to make a point: because we don't have access to other people's qualia, we can't tell what they are experiencing (or not experiencing, as in the case of a philosophical zombie).

For students unconvinced by the zombie exercise (because, after all, philosophical zombies are just a figment of a philosopher's imagination), I have them do a chocolate exercise.

Everyone receives a piece of chocolate, and the task is for one student to taste the chocolate and describe his or her taste experience to a partner. To make things more interesting, I tell them that their partner has never tasted chocolate. Students often say things like "it tastes sweet," which raises the question of what the experience of sweet is like, or "I like it a lot," which is only tangentially related to the actual taste experience. You get the point: even real people who have all experienced qualia cannot transmit the *essence* of their experience to another person. We therefore conclude that *perceptual experience is private*. We know we are having an experience, and we may think other people are having similar experiences, but the only thing we can be sure of is our own experience.

Can Experience (Qualia) Be Inferred from Scientific Knowledge?

The idea that we have no way of knowing what qualia another person is experiencing led the philosopher Frank Jackson to propose a thought experiment called "Mary the Color Scientist."[39]

Mary, according to Jackson's thought experiment, was raised from birth in a black-and-white world. Chromatic colors such as blue, red, and green were not part of her world. Everything was white, black, or gray. While she was growing up in this colorless world, Mary studied the science of chromatic color, until eventually she knew everything there was to know about the psychological and physiological processes responsible for creating colors like blues, reds, and greens.

Finally, Mary is released from her color-deprived environment and sees chromatic colors for the first time. How will she react when she sees a red rose? Will she access her scientific knowledge base and say, "Yes, I knew it would look like that"? Or will she be amazed and say, "Wow! So that's what red looks like. I never would have guessed"? According to Jackson (and most of my students, when I pose this problem to them), Mary would be surprised. Scientific knowledge can only go so far and thus cannot be used to predict qualia. Experience, as it turns out, needs to be experienced.

This thought experiment would not be relevant to our quest for understanding qualia except that a real "Mary" has recently been discovered. Her name is Dr. Susan Barry, and she is a professor of neurobiology at Mount Holyoke College.

One of the topics Susan teaches in her neurobiology course is how depth perception results from binocular vision—vision that involves both the left and right eyes. Because each eye sees the world from a slightly different vantage point, we register slight differences between the images of the environment in the left and right eyes. (Hold a finger about one foot away. Look at it and blink back and forth between the left and right eye and notice how the scene behind the finger moves back and forth.)

Although there are differences between the images in the two eyes, they are similar enough that the brain can combine them into a single image—a process called *fusion*. The slight differences between the images in the two eyes are, however, registered by the nervous system and provide information about the depths of objects in a scene. This information results in perception of depth in a scene, which is called *binocular depth perception*.[40]

However, although Susan told the story of binocular vision to her class, she had never experienced it herself. She explained this to the famous neurologist Oliver Sacks when she met him at a party. When she was little, she said, she had a condition called *strabismus*, in which one of her eyes would look in one direction and the other a little off to the side. Because the left- and right-eyed images could not be matched up and fused, her visual system suppressed vision in one eye so that she would not experience double vision. Because of this suppression, Susan saw the world through one eye, a situation called *monocular vision*.

As a girl, Susan had an operation designed to realign her eyes, but her monocular vision continued. Because of this, she did not experience binocular vision or binocular depth perception. Susan could still judge depth by using cues like overlap (Sam is partially covering Charlie, so Sam must be closer), but she lacked the enhanced depth perception created by binocular vision. You can appreciate the difference between monocular and binocular depth perception by comparing your perception of depth you experience when viewing regular movies, which are projected onto a flat or slightly curved screen, and the perception of a 3-D movie in which special glasses are used to achieve binocular depth perception.

At the party, as Susan was explaining her situation to Dr. Sacks, he asked her, "Can you imagine what the world would look like if viewed with two eyes?" She replied that yes, she could imagine it, because, after all, as a professor of neurobiology, she had read hundreds of papers on binocular depth perception. But in December 2004, almost nine years after their conversation, Susan wrote a letter to Sacks, which began, "You asked me to imagine what the world would look like when viewed with two eyes. I told you I thought I could...but I was wrong." Susan had gone to an optometrist, who prescribed eye exercises designed to make her eyes work together. These exercises changed her perception, and her book *Fixing My Gaze* describes what started happening as her eyes began working together:[41]

> The sun was setting as I left Dr. Ruggiero's office.... I got into my car, and as I looked up to adjust the rearview mirror, the mirror popped out at me floating in front of the windshield. I was transfixed.... A

large sink faucet reached out toward me, and I thought I had never seen such a lovely arc as the arc of the faucet. The grape in my lunchtime salad was rounder and more solid than any grape I had ever seen before. I could see, not just infer, the volume of space between tree limbs, and I loved looking at, and even immersing myself in, those inviting pockets of space.

So Susan had become the "Mary" of depth perception. Despite all her scientific knowledge and her confidence that she knew what the experience of depth perception would be like, she was nonetheless amazed when she first experienced binocular depth perception. Scientific knowledge was not enough. She needed to actually *experience* binocular depth perception, not just know about it.

In addition to the idea that experience must be experienced, we can draw another message from Susan's story. After Susan's eyes became lined up, she experienced binocular perception when neural signals sent from her left and right eyes met and were compared in her brain. Think about what this means. Nerve firing caused by the two-dimensional image of the scene in the left eye met up with firing caused by the two-dimensional image of the scene in the right eye, and Susan experienced three-dimensional vision. Somehow the brain had transformed two flat images into a single three-dimensional perception. What this means is that activity in the nervous system *creates* experiences.

The Creation of Experience

To understand what it means to say that the nervous system creates experiences, let's consider two scenarios about perceiving the world:

Scenario 1: There are things out there in your environment. A red balloon floating high in the blue sky catches your attention.

If your senses are functioning normally, the receptors in your eyes and the activity in your brain will determine that there is a red balloon in the sky, which is seen against a background of the blue sky. In other words, your experience will indicate what is out there.

This is a nice story, but it is only partially true. The next scenario is closer to the truth.

Scenario 2: You are in the same scene as before. Receptor cells in your eye detect a round object above that reflects long-wavelength light into your eyes. The sky in the background sends short-wavelength light into your eyes.

The second scenario is more accurate than the first because it describes the environment in terms of the energy reaching your visual receptors. Consider the balloon. Is it really red? No. It reflects long-wavelength light, and your visual system turns that light into the perception "red." The red is in your mind, not in the balloon. The same is true for the sky. It sends short-wavelength light into your eyes, and your visual system turns this light energy into the perception "blue."

"Okay," you say, "I get the part about rays of light entering the eye and causing the perception of colors. But doesn't the long-wavelength light reflected from the balloon simply indicate that the balloon is red?" If you were saying that to Isaac Newton, he would disagree. Here is what Newton said about light rays back in 1704:

> The Rays to speak properly are not coloured. In them there is nothing else than a certain Power and Disposition to stir up a Sensation of this or that Colour.... So Colours in the Object are nothing but a Disposition to reflect this or that sort of Rays more copiously than the rest.[42]

Newton's idea is that the color you see in response to the light reflected from the balloon is not contained in the rays of light themselves, but the rays "stir up a sensation of that color." Stating this idea in modern-day physiological terms, we would say that light rays are simply energy, so there is nothing intrinsically "blue" about short wavelengths or "red" about long wavelengths, and *we perceive color because of the way our nervous system responds to this energy.*

Experience Is Created by the Nervous System

An indication that colors are indeed created by receptors is that many animals perceive either no color or a greatly reduced palette of colors compared to humans, and others sense a wider range of colors than humans, depending on the nature of their visual systems. For example, honeybees have a visual receptor that responds to very short wavelengths that humans can't perceive.[43] What "color" do you think bees perceive at these wavelengths, which you can't see? You might be tempted to say "blue" because humans see blue at the short-wavelength end of the spectrum, but you really have no way of knowing what the honeybee is seeing, because, as Newton stated, "The Rays...are not coloured." There is no color in the wavelengths, so the bee's nervous system *creates* the bee's experience of color. For all we know, the honeybee's experience of color at short wavelengths is quite different from ours and may also be different for wavelengths in the middle of the spectrum that humans and honeybees can both see.

The idea that the nervous system is responsible for the quality of our experience also holds for other senses. For example, our experience of hearing is caused by pressure changes in the air. But why do we perceive slow pressure changes as low pitches (like the sound of a tuba) and rapid pressure changes as high pitches (like a piccolo)? Is there anything intrinsically "high-pitched" about rapid pressure changes? Or consider the sense of taste. We perceive some substances as "bitter" and others as "sweet," but where is the "bitterness" or "sweetness" in the molecular structure of the substances that enter the mouth? The answer to this question is that our sensory experiences may be *triggered* by light entering the eye, pressure changes within the ear, or the structure of molecules in the mouth, but the actual experience is a creation of our nervous system.

How Is Experience Created by the Nervous System?

> Consciousness poses the most baffling problems in the science of the mind. There is nothing that we know more intimately than conscious experience, but there is nothing that is harder to explain.
> —David Chalmers[44]

It is April 12, 1994, in a lecture hall on the campus of the University of Arizona, and the philosopher David Chalmers steps up to the podium to present his paper at the first Tucson consciousness conference, titled "Toward a Scientific Basis for Consciousness." The audience is restless, bored by the previous talks, and looking forward to the coffee break that follows directly after Chalmers's talk.[45] Chalmers, however, gets their attention when he introduces the idea of the *hard problem of consciousness*—the problem of determining how physical processes are transformed into experience.

What does it mean to say "how physical processes become transformed into experience"? One way to answer this question is to consider *transduction*, the process that occurs when stimuli in the environment are translated into electrical signals in the nervous system. For example, in vision, some of the light that enters the eye hits the retina at the back of the eye, finds its way into visual receptors within the retina, and is absorbed by chemicals within the receptors called visual pigments. These pigments change their shape, in reaction to the light, and this triggers a complicated sequence of chemical events that culminates in the generation of electrical signals, first in the receptors, and then in other neurons in the retina. These signals travel out of the back of the eye in the optic nerve and eventually reach the brain, where our experience is created.[46] So the question Chalmers is posing is "how do electrical signals in the brain become transformed into experience?"

But to really appreciate the hard problem, we need to look at how these electrical signals are created. To do this, let's zoom in on one place along a nerve fiber and observe what happens when an electrical signal traveling down the fiber passes by. The first thing we notice is that the nerve fiber is surrounded by water (picture a long, hollow tube underwater), which contains charged molecules called ions (fig. 2.2a). Two of the major ions are positively charged sodium (Na+) and potassium (K+). As an electrical signal passes our point of observation, we see Na+ ions flowing from the outside of the fiber to the inside (fig. 2.2b), followed by K+ ions flowing from the inside to the outside (fig. 2.2c). Many other things are happening as well, but these two ion flows cause the inside of the fiber to become positive for a fraction of a second, creating a positive wave called the *nerve impulse*, which travels down the fiber. Nerve impulses are therefore "wet" electrical signals caused by ions flowing across membranes. With this knowledge about the chemical nature of nerve impulses, we can now state the hard problem as follows: "How is the flow of Na+ and K+ across the nerve membrane transformed into experience?"

When the problem is stated in this way, we can appreciate why Chalmers coined the term "*hard* problem." Although we know a lot about transduction from physical energy into electrical energy, and how the flow of chemicals create nerve impulses,

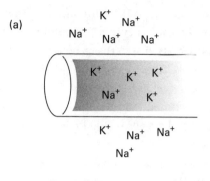

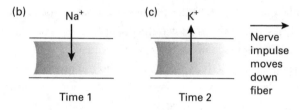

Figure 2.2
(a) Section of a nerve fiber showing positively charged sodium (Na) and potassium (K) ions inside and outside the neuron. More sodium ions are present outside and more potassium ions inside. (b) Watching ion flow at one place along the nerve fiber, you see sodium flowing in, followed by (c) potassium flowing out. This flow creates the nerve impulse, which travels down the fiber.

we still don't know how these physical events are transformed into experience.

After Chalmers's talk, the previously somnolent audience woke up, and during the coffee break, the hard problem became the number one topic of conversation.[47] It is interesting that the term "hard problem" caused such a sensation because, as Chalmers is the first to admit, "hard problem" is just a catchy phrase for something people knew all along.[48] Consider, for example, the following statement by the English biologist Thomas Huxley (1825–1895): "How is it that anything as remarkable as a state of consciousness comes about as a result of irritating nervous tissue is just as unanswerable as the appearance of Djina (the genie) when Aladdin rubbed his lamp."[49] Huxley equates the creation of a state of consciousness by nervous tissue to the appearance of a genie from Aladdin's lamp. What could cause such a thing? Magic? God? No one knows. Our inability to explain how the jump from charged molecules flowing across a membrane to experience occurs was described by the philosopher Joseph Levine (1983) as the *explanatory gap*. Jumping the chasm between *physiology* and *experience* is a very hard problem indeed and involves a gap that science has not yet been able to explain.[50]

It is easy to see why, in Huxley's day, limitations in knowledge of physiology would make it difficult to study the hard problem. But what about present-day scientists, who are equipped with sophisticated technology for measuring physiological responses of the brain?

As it turns out, technology has led to a great deal of research on the brain, but no solution to the hard problem. If we consider the data that technology provides, it is easy to understand why this is so. Consider, for example, an experiment by William Newsome and coworkers, in which they recorded from single neurons in a location in the monkey's cortex called the medial temporal (MT) area and found that neurons is this area are activated by dots moving in a specific direction (see fig. 2.1).[51] This finding locates an area in the brain that is involved in perceiving motion. But the relationship between nerve firing and motion perception is *correlational*—that is, it tells us that motion perception is *related* to the firing of MT neurons, but it does not tell us how the firing of neurons is *transformed* into our perception of motion.

Understanding the transformation is the hard problem, and though we now possess the technology to record nerve impulses, technology does not solve the hard problem.

Here is another reaction to the hard problem, by the developmental psychologist Alison Gopnik, who calls the hard problem the "capital C question":

> Now, of course, there's always been this big question... the capital C question of consciousness. How can a brain have experiences? I'm skeptical about whether we're ever going to get a single answer to the big capital C question. But there are lots of very specific things to say about how particular kinds of consciousness are connected to particular kinds of functional or neural processes.[52]

In concluding here that there are specific things to say about how consciousness is connected to neural processes, Gopnik is basically saying, "Okay, we can't solve the hard problem, so let's look for connections between consciousness and neural processes." Establishing these connections, which are called *neural correlates of consciousness* (NCC), was labeled by Chalmers as "the easy problem."

The Neural Correlates of Consciousness

A person lies in a brain scanner in Jack Gallant's laboratory at the University of California, Berkeley.[53] His brain activity, measured using functional magnetic resonance imaging, is recorded as he looks at a photograph of a living room containing some furniture. This activity is displayed as a pattern of activity in units called voxels, where each voxel is a small volume of the brain (see fig. 1.3e).

The magic happens when the pattern of voxel activity caused by the photograph of the living room is entered into the decoder. This decoder has previously been calibrated by having a number of people look at thousands of pictures, each of which creates a pattern of voxel activity that is entered into the decoder. Each voxel pattern is linked to the objects and scenes that generated it, so the decoder eventually "learns" what patterns of activity go with specific types of objects and scenes.

The pattern of activity generated by the picture the person is looking at is analyzed by the decoder, and even though this picture has never been "seen" by the decoder, it comes up with an answer: "living room." This process is repeated for other types of scenes, like athletes running on a track, a fish in an aquarium, and a city waterfront with boats. The decoder comes up with an answer for each scene: "people moving," "water animals," "urban street/boatway." The decoder's answers are not right every time, but it performs significantly better than chance.

This experiment and others show that it is possible to read a person's mind by analyzing his or her brain activity.[54] Remember that the person sees the picture, but all the decoder sees is the brain activity caused by the picture, and it determines what the person was seeing based on this brain activity. The principle at work here is that each scene generates a different pattern of brain activity, and the decoder recognizes the type of scene based on this pattern.

Everything you look at, think about, or remember creates a pattern of brain activity that your mind miraculously transforms into perceptions, thoughts, or memories. As we saw in our discussion of the "hard problem," we still don't know how patterns of electrical activity are *transformed* into experiences. But we do know that different patterns of activity *represent* different experiences, and this is the principle behind both the operation of Jack Gallant's decoder and the operation of your mind.

If you were able to view what is happening in a person's brain during routine daily life, you would observe what has been described as a symphony of electrical activity throughout the brain. And if you tried to understand this symphony, you would be overwhelmed by how complex and widespread it is. But this symphony of activity isn't random. It's organized so that each of your mental experiences—memories, perceptions, thoughts—are associated with a particular pattern of activity.

One route to understanding how the brain's activity is organized is to determine the functions of specific structures. Researchers have done this using the techniques of neuropsychology, electrophysiology, and brain imaging (see chap. 1). Research in the nineteenth century through the mid-twentieth century identified "primary" areas for the senses of vision, hearing, and the skin senses, as well as areas involved in memory and emotion. This linking of functions to specific structures is called *localization of function*, and this is where our description of brain organization begins.

Localization of Function

> The entire brain system can be decomposed into subsystems or modules.
> —Danielle Bassett and Michael Gazzaniga[55]

Localization of function is the idea that different areas of the brain are specialized to do specific things. This was not obvious to early researchers, who pictured the cortex as a homogeneous mass, with all functions being handled by the brain as a whole, but in the mid-1800s Paul Broca and Carl Wernicke identified an area in the frontal lobe (Broca's area) responsible for language production, and an area in the temporal lobe (Wernicke's area) responsible for language comprehension (see fig 1.3a). The idea of localization of function was extended to vision by showing that removing the occipital lobe in monkeys caused blindness, and patients with occipital lobe damage caused by stroke also became blind.[56] The occipital lobe is therefore also referred to as the *primary visual receiving area* or simply the *visual cortex* (see fig. 2.1).

Functional organization was also discovered in subcortical structures such as the *hippocampus* and the *amygdala*, located beneath the cerebral cortex. The hippocampus was identified as being essential for storing memories in 1953, when Henry Molaison underwent an experimental procedure to eliminate severe epileptic seizures. The procedure, which involved removing his hippocampus on both sides of his brain, succeeded in decreasing his seizures but had the unintended effect of eliminating his ability to form new memories.[57] One result of this inability to form new memories was that even though the psychologist Brenda Milner tested him many times over many decades, Henry always reacted to her arrival in his room as if he were meeting her for the first time.

The amygdala, near the tip of the hippocampus, was linked to emotions by cases in which damage to the amygdala caused problems in recognizing and feeling emotions. For example, patient SM, whose amygdala was completely destroyed on both sides of her brain by an inherited disease, experienced reduced emotion in her life and also responded differently than normal to emotional stimuli.[58] When she rated the amount of "arousal" caused by pictures of emotional faces, she assigned much lower than normal arousal ratings to negative emotions like fear and anger. An angry face that was rated 7.4 out of 9.0 for arousal by normal participants was rated 1.2 by SM.[59]

Perhaps even more significant was SM's reaction to emotionally charged sentences. For example, her arousal rating for the sentence "Sally waved her hands in the air and yelled for help, as the boat was sinking," was 3.0. Normal participants gave this sentence a 9.0, the highest rating possible. The study of SM, plus research showing that in non-brain-damaged participants the amygdala responds more vigorously to fearful faces than to happy faces, has led to the conclusion that the amygdala plays an important role in processing negative emotions.[60]

More than a century of research on localization of function has left no doubt that both cortical and subcortical brain areas are specialized to handle specific functions. But the idea of one structure controlling one function turned out to be oversimplified. One hint that things are more complex than simply "vision is located in the occipital cortex" was provided by researchers who recorded from single neurons in the occipital cortex and asked, "What type of visual stimulus causes the best response in this neuron?" Two early researchers who adopted this strategy were David Hubel and Torsten Wiesel, who won the Nobel Prize in 1981 for their research on the visual system.[61]

Hubel and Wiesel found that some neurons in the occipital cortex, where the signals that had been transmitted from the eye were first arriving, responded best to small spots of light. Probing deeper into the visual cortex, they discovered other neurons that responded to lines or bars of a particular orientation and some that responded only to bars that were moving in a particular direction.

Hubel and Wiesel's discovery of these neurons showed that neurons at each successive stage in the visual system become more specialized and so respond to more complex stimuli. This idea was confirmed by a new generation of researchers who, following Hubel and Wiesel's lead, began determining the "best stimulus" for neurons outside the occipital cortex and found neurons at higher levels that responded to complex shapes, faces, bodies, parts of bodies, rooms, and buildings.[62]

As it turns out, about 25 percent of the human cerebral cortex is predominantly visual in function.[63] With the realization that vision stretches across the cortex, we need another concept to explain cortical organization. That concept is *distributed representation*.

Distributed Representation

Distributed representation is the idea that a particular mental function is represented by activity in areas that are distributed throughout the brain. If this is true, you might wonder whether localization of function is still valid. The answer is yes, because areas that are specialized for different functions are connected together to create activity distributed throughout the brain.

It is also important to note that localization is not absolute, because although an area called the *fusiform face area* (FFA) responds strongly to faces, it also responds to other things, like other types of objects, scenes, cars, and birds. Moreover, neurons outside the FFA also respond to faces.[64] This means that the FFA is strongly activated if you look at a face, but other areas are activated as well.

The distributed neural response to faces reflects the fact that our experience with faces extends beyond simply identifying an object as a face ("that's a face"). We can also respond to the following additional aspects of faces: (1) emotional aspects ("she is smiling, so she is probably happy," "looking at his face makes me happy"); (2) where someone is looking ("she's looking at me"); (3) how parts of the face move ("I can understand him better by watching his lips move"); (4) how attractive a face is ("he has a handsome face"); and (5) whether the face is familiar ("I remember her from somewhere"). This multidimensional response to faces is reflected in distributed neural responses throughout the cortex.

If looking at a face can activate many areas, just imagine the activity caused by looking at a visual scene that might contain faces, plus many other objects as well. A typical visual scene includes visual qualities such as shapes, colors, textures, light and shadows, movement, and spatial relations between objects, each of which activates different brain areas. And to make things even more interesting, different kinds of objects (nonanimals, like tools and furniture, versus animals, like cats and dogs) activate different locations in the brain.[65] It's no wonder, then, that looking at a face or a scene creates the symphony of activity described at the beginning of the chapter.

I have been focusing on vision because we know so much about it. But the mind is about much more than seeing, and distributed representation occurs for other senses, like touch, smell, and pain, and for mental functions such as attention, memory, emotions, and dealing with social situations. Memories, for example, are complicated. Some memories—called short-term memories—last fleetingly, for only about fifteen to twenty seconds unless repeated over and over, as you might do to memorize a phone number you forgot to store in your phone. Other memories are longer, such as your memory for something you did last week or even years ago. Henry Molaison, who lost the ability to form new memories when his hippocampus was removed could still remember things that had happened within the most recent fifteen to twenty seconds. This means that he still retained short-term memory, and short-term memory must be served by a different area of the brain from the hippocampus.[66] But memories can also differ in another way. *Episodic memories* are memories for events in a person's life, like remembering what you did yesterday. *Semantic memories* are memories for facts, like knowing that the capital of California is Sacramento. Brain-scanning experiments indicate that thinking about episodic and semantic memories activates different areas of the brain.[67]

In addition, remembering activates areas throughout the brain for other reasons. Memories can be visual (picturing someplace you often visit), auditory (remembering a favorite song), or olfactory (smell triggering memories for a familiar place). Memories often have emotional components, both good and bad (thinking about someone you miss). Most memories combine many of these components, each of which activates different areas of the brain. Memories, like faces, create their own symphony of neural activity.

Early in the first chapter, I noted that we could describe the scientific approach to the study of the mind as "a long and winding road." We have, in these first two chapters, been traveling along that road, with some side trips along the way to consider things like the mind in popular culture, skepticism regarding whether a connection exists between mind and brain, musings regarding zombies, and whether we can know what a bat, or even your best friend, is experiencing.

But whatever side trips we took, we always came back to the idea of connections between the mind and the brain. Chapter 1 ended by introducing some of the procedures used for measuring mind-brain connections (see table 1.1), and this chapter has ended by introducing some basic principles of brain organization.

Now, as we move on to chapter 3, we leave unscientific speculation and the hard problem of consciousness behind to devote all our energy to describing what physiological and behavioral research has revealed about how the mind works. To begin this discussion, we consider how creation of the mind involves mechanisms that are largely hidden from view.

3 The Hidden Mind

> Everything we see is a shadow cast by that which we don't see.
> —Martin Luther King Jr.

As I write this, sitting outside on my porch, surrounded by tall trees in my backyard, I can hear birds singing and chirping, greeting the morning. But as I look out, all I can see are the trees. I know the birds are out there because I can hear them, but they are hidden, out of view.

Are these invisible tweeting birds trying to tell me something? They are, after all, creating my experience while remaining invisible, which is exactly what the mind does. You know your mind is working because it enables you to read these words, but as you read, you aren't aware of the tens of thousands of nerve impulses that are coursing along highways of nerve fibers in your brain or the grammatical rules that you are employing to combine the words into meaningful sentences. You know your mind is working because when you look up, you see the scene around you. But you aren't aware that your ability to make sense of this scene depends on knowledge that has been stored in your brain through countless earlier experiences of seeing scenes.

Somehow, this internalized knowledge transforms all the individual elements of your environment into your perception of a coherent scene.

What does it mean to say that the mind is "invisible"? One way to answer that question is to state the obvious fact that we are unaware of the operation of mental processes such as the decisions studied by Donders in his reaction-time experiment (see chap. 1), physiological processes such as nerve impulses and brain activations (chap. 1), and the physiological processes that create consciousness (chap. 2). This chapter continues our discussion of hidden processes by first looking at research that has revealed fascinating things about the physiology of the mind by studying patients who have suffered brain damage.

Hidden Processes Revealed by Brain Damage

Although devastating to the person who experiences it, brain damage has enabled researchers to study mechanisms that are normally hidden from view. We will see what this means by first considering a phenomenon called *visual form agnosia*.

Visual Form Agnosia: Seeing without Recognizing

Continuing in the tradition of Broca and Wernicke, who determined areas in the frontal and parietal lobes that are important for understanding and processing speech, the neurologist Oliver Sacks described the case of Dr. P. in the title story of his book *The Man Who Mistook His Wife for a Hat*.[1]

Dr. P., a well-known musician and music teacher, first noticed a problem when he began having trouble recognizing his students visually, although he could immediately identify them by the sound of their voices. But when he began misperceiving common objects, for example, addressing a parking meter as if it were a person or expecting a carved knob on a piece of furniture to engage him in conversation, it became clear that his problem was more serious than just a little forgetfulness. Was he blind, or perhaps crazy? It was clear from an eye examination that he could see well, and by many other criteria, it was obvious that he was not crazy.

Dr. P.'s problem was eventually diagnosed as *visual form agnosia*—an inability to recognize objects. His particular agnosia was caused by a tumor that affected visual processing in his occipital and parietal lobes. He perceived individual parts of objects but could not integrate the parts into a whole object. Thus, when Sacks showed him pictures of scenes in a *National Geographic* magazine, Dr. P. was able to identify details but could not create a scene-as-a-whole in his mind. When Sacks showed him a rose, Dr. P. described it as "a convoluted red form with a linear green attachment." When Sacks asked him what it was, Dr. P. said that "it is not easy to tell" but guessed that "it could be a flower." When he was offered the opportunity to smell the object, he immediately identified it as a rose. Thus he knew what a rose was, but he could not construct one visually. Patients with visual form agnosia who can't identify objects visually can also typically close their eyes and identify objects based on touch alone.[2]

How does visual form agnosia relate to the idea that there are hidden processes in the brain? One thing it tells us is that perceiving involves *neural processing*; it's not just a matter of signals associated with different parts of an object reaching the visual area. The representation of each part of an object or scene must be combined to create whole objects and coherent meaningful scenes. The mechanism that accomplishes this does not function in people who have visual agnosia.

Let's take the idea of dealing with objects in the real world a step farther by considering the case of DF, a thirty-four-year-old woman who collapsed and lost consciousness while taking a shower because of carbon monoxide poisoning from a leak in a faulty propane heater.[3] Although DF was initially blind, over time she regained some vision, being able to name bright colors, but she described her vision as "blurred" and scored poorly on tests of shape perception and detail vision. She was also unable to identify objects, for example, identifying a screwdriver as "long, black, thin," and was diagnosed as having visual form agnosia.

But DF wasn't completely incapacitated. She was able to do tasks that involved physical actions such as avoiding furniture as she walked through a room, shaking people's hands, and opening doors. These abilities led David Milner and coworkers to present DF with a task to test her ability to take action by "mailing" a card through an oriented slot (fig. 3.1a).[4] DF grasped the card, and as she started moving it toward the slot, she rotated it to match the slot's orientation and slipped the card through the slot.

The feat of mailing the card might not seem particularly noteworthy, except that when DF was given the task of simply rotating the card so that it matched the orientation of the slot, without actually moving the card toward the slot, she was unable to do so (fig. 3.1b). Thus DF performed poorly in the *static* orientation-matching task but did well when *action* was involved. DF's behavior shows that we have one mechanism for judging orientation and another for coordinating vision and action.[5]

One way to understand DF's behavior is to consider the results of experiments on monkeys, which showed that there are two pathways that send signals from the visual cortex to different areas of the brain (fig. 3.2).[6] The *ventral pathway*, which sends signals from the occipital lobe to the temporal lobe, on the lower

The Hidden Mind

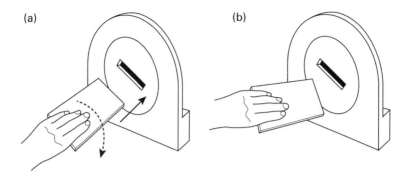

Figure 3.1
(a) DF, who suffered damage to the temporal lobe of her brain, was able to "mail" a card through a slot by rotating it to the correct orientation as she moved it toward the slot. (b) However, she couldn't perform the static orientation task, which involved matching the orientation of the card to the slot without moving the card toward the slot.

side of the brain, is responsible for identifying objects. This pathway is therefore called the *what* pathway. Removing a monkey's temporal lobe makes it unable to perform a task that involves telling the difference between two differently shaped objects. DF's damage was in the temporal lobe, so her *what* pathway was not available for identifying objects or matching orientations.

The *dorsal pathway*, which sends signals from the occipital lobe to the parietal lobe, at the top of the brain, is responsible for controlling actions toward objects and is called the *how* pathway, with some researchers calling it the *where* pathway. Removing a monkey's parietal lobe makes it unable to do a task that involves responding to objects in different locations.[7] Because DF's *how* pathway was intact, she was able to "mail" the card.

The distinction between *what* and *how* pathways illustrates hidden processes in two ways. First, we are unaware that a

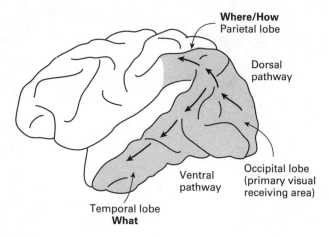

Figure 3.2
The monkey cortex, showing the *ventral*, or *what* pathway from the occipital lobe to the temporal lobe, and the *dorsal*, or *where/how pathway*, from the occipital lobe to the parietal lobe.

simple action like mailing a letter involves complex interactions between two widely separated areas of the cortex. Second, and perhaps most important, the actions controlled by the *how* pathway do not require conscious knowledge.[8] Thus, if you were to observe DF while she was being tested, you would see her, in the static card condition, try, in vain, to adjust the card's orientation to match the slot. But then, miraculously, as soon as she starts moving the card toward the slot, it rotates to match the slot's orientation. I say "it" rotates because DF isn't consciously rotating the card. Her intact *how* pathway is automatically controlling her actions without her awareness. V. S. Ramachandran and Sandra Blakeslee, in their aptly titled book *Phantoms in the Brain*, describe DF's actions as occurring "as if some other person—an unconscious zombie inside her—had guided her actions," and she had moved the card "without any conscious awareness, as if that very same zombie had taken charge of the task and effortlessly steered her hand toward the goal."[9]

DF's disabled *what* pathway reveals the functioning of her unconsciously controlled *how* pathway. But what is important about this is that, in people without brain damage, both of these pathways operate in coordination with each other. Consider, for example, what happens as you reach across a cluttered table to pick up a cup of coffee. You begin by observing the table and identifying the objects on it with your *what* pathway, which identifies the cup of coffee. You then begin reaching for the cup, guided by your *how* pathway, carefully avoiding nearby objects that you've identified with your *what* pathway.

Looking at the cup, your *what* pathway notes the shape of the cup's handle, just in time for your *how* pathway to adjust your hand's final trajectory, adjusting the spacing of your fingers to grasp the handle, and finally adjusting the force needed to pick up the cup, taking into account information your *what* pathway has provided about how much coffee is in the cup. All this drama back and forth between two widely separated pathways in the brain just to pick up a cup of coffee!

Blindsight: Detecting without Seeing

Another case of damage to one area of the brain highlighting the functioning of another area has occurred in some rare side effects of occipital lobe damage. George Riddoch, while studying British soldiers who had been wounded in World War I, found that damage to certain areas of the occipital cortex caused blindness in specific areas of the visual field.[10] But he also observed a few cases of soldiers who were blind in a particular area of space but could still detect motion in that area. When Riddoch held up a finger in the blind area of the visual field, the soldier could not see it. But when he moved the finger, the soldier reported that "something was moving." The soldier was not sure what that "something" was or if it even had a form, but he could tell that it was moving.

Over fifty years later, Lawrence Weiskrantz, studying a patient called DB, extended Riddoch's observation by making a related discovery regarding "perception" in the blind area of the visual field.[11] DB was a thirty-four-year-old man who had an operation in which a portion of his occipital lobe was removed to eliminate a tumor that was causing headaches and hallucinations. When DB's vision was tested by visual perimetry—a technique in which people are asked to indicate when they see spots of light flashed in different locations in their visual field—Weiskrantz found that DB could see on the right side of his visual field but was blind to lights flashed on the left.

Once the locations of the seeing and blind areas were determined, Weiskrantz made what DB probably thought was a strange request. "I'm going to flash a spot of light in your blind field, and I want you to point with your finger where the spot was located." Remarkably, DB accurately located light flashes that appeared in his totally blind left visual field. DB was amazed because he said he had not seen anything and was just guessing. Weiskrantz called DB's ability to detect the location of a stimulus he could not see *blindsight*.

DB could also indicate whether outstretched fingers in the blind area were moving vertically or horizontally (when given a choice between the two) or whether an *X* or an *O* had been presented in his blind area. As in the case of the light flashes, DB claimed he had not seen anything, either when judging the direction of movement or whether there was an *X* or an *O*.

This ability to somehow detect things even in blind areas of the visual field is an example of a hidden process revealed by brain damage. But how is this possible? One idea is that blindsight reveals the operation of a visual pathway in addition to the main pathway from the eye to the occipital cortex.[12] Although most of the signals sent from the eye reach the visual cortex, about 10 percent of these signals reach a subcortical structure called the superior colliculus (SC). Signals from the SC are then sent to structures in the *how* pathway, which serve normal motion perception. The mystery, yet to be solved, is why an area that is involved in normal motion perception can, when the occipital cortex is missing, create the mysterious ability to answer questions about unseen motions and objects.

Visual Neglect: What's Happening Off to the Side?

Another mystery involving a loss of awareness is posed by a disorder called *visual neglect*, which often occurs after damage to areas near the border between the temporal and parietal lobes in the right hemisphere of the brain. People with visual neglect due to right hemisphere damage behave as if the left half of their visual world no longer exists.[13] They ignore people and objects to their left, may eat food from only the right side of their plate, or shave only the right side of their face. When asked to copy a picture of an object, they often omit details from the left side of the object.

One conclusion from symptoms such as these might be that the person is blind on the left side of the visual field. But that can't be the case, because patients with visual neglect *do* see things on their left if they are told to pay attention to what is on the left side of the environment. The problem therefore seems to be a lack of *attention* to what is on the left, rather than a loss of *vision* on the left.

We can gain more insight into visual neglect by testing patients in the laboratory. If they are told to look at a "fixation cross," so that they are always looking straight ahead, and a light is flashed to their left side, they report seeing the light. This is another indication that they are not blind on their left side. But if two lights are simultaneously flashed, one on the left and one on the right, patients report that they see a light on the right, but they do not report seeing a light on the left. This loss of awareness on the left when a competing stimulus is presented on the right is called *extinction* and occurs in some patients with visual neglect.

Extinction provides insight into visual neglect because it suggests that the neglect of objects in the environment is caused by competition from the stimuli on the right, with the left ending up being the loser. This is another example of hidden processes, because when stimuli occur only on the left, the signals generated by these stimuli are transmitted to the brain, and the patients see the stimulus. However, when a stimulus is added on the right, the same signal is still generated on the left, but the patients don't "see" on the left because their attention has been distracted by the more powerful stimulus on the right.

"Seeing" or "conscious awareness" is therefore a combination of signals sent from stimuli to the brain *and* attention toward these stimuli. This happens in non-brain-damaged people as well, who are often unaware of things happening off to the side of where they are paying attention—but there is a difference. Although they may miss things that are off to the side, they are aware that there is an "off to the side"!

We still have more to learn from the phenomenon of neglect and extinction, because it turns out that extinction can be partially eliminated for certain types of stimuli. When a ring stimulus

The Hidden Mind

that the patient has never seen before is presented on the left, and a flower stimulus that has also not been previously seen is presented on the right, patients with visual neglect see the ring in only 12 percent of the trials (fig. 3.3a). Extinction is therefore high when the ring is on the left. However, when the stimuli are switched so that the flower appears on the left, perception rises to 35 percent (fig. 3.3b). Finally, when a spider is presented on the left, it is seen in 80 percent of the trials (fig. 3.3c).[14] In a similar experiment, patients were more likely to see sad or smiley faces on the left than neutral faces.[15]

Why is the patient more likely to see the spider? The answer seems obvious: the spider attracts attention, perhaps because it is menacing and causes an emotional response, whereas the flower shape does not. But how do patients know that the shape on the left is a spider or a flower? They need to have seen the spider or flower, right? But if they have seen them, why do they, as in the case of the ring and the flower, often fail to report seeing them?

What apparently is happening is that the flower and spider are processed by the brain at a subconscious level to determine their identity, and then another process determines which stimuli will be selected for conscious vision. The identification at a subconscious level is called *preattentive processing* to indicate that processing is occurring in the visual system *before* attention is assigned. The patient is not aware of preattentive processing, because it is hidden and happens within a fraction of a second. What the patient is aware of is which stimuli are selected to receive the attention that leads to conscious vision.[16]

In all the cases above, brain damage causes a loss of some kind, and this loss reveals processes that are invisible or difficult to detect in the normally functioning brain. But we also have

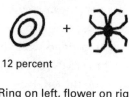

12 percent

(a)   Ring on left, flower on right

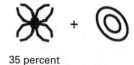

35 percent

(b)   Flower on left, ring on right

78 percent

(c)   Spider on left, ring on right

Figure 3.3
A series of tests to determine the degree of extinction for different pairs of stimuli. The number below the left image indicates the percentage of trials in which the image was identified by a patient who usually shows neglect of objects in the left side of the visual field. (a) Ring on left, flower on right; (b) flower on left, ring on right; (c) spider on left, ring on right.

ways of discovering hidden processes in the undamaged brain. One example, which we will consider next, involves recording electrical signals that are generated as a person makes a decision.

What Happens in the Brain Just Before We Make A Decision?

Remember Donders, who was interested in determining how long it took to decide whether a light was presented on the left or on the right (chap. 1)? We now return to decisions, but from a different angle, as we look at research that asks what happens during the time just before we are making a decision to move part of our body.

Benjamin Libet's Famous Experiment: When Does a Decision Begin?

The question of what happens in the brain just before we make a decision was posed by Benjamin Libet and coworkers, in an experiment published in 1983 that has been described as "possibly the most important psychological experiment ever."[17] Although not every researcher may agree with this statement, it is not an understatement to say that Libet's finding—that there is an electrical signal in the brain that occurs *before* a person makes a decision—has generated a huge amount of discussion among researchers.

Libet's participants looked at a clock about two meters away (fig. 3.4a). They were fitted with scalp electrodes, which measured the electroencephalogram (EEG), the electrical response of thousands of neurons under each electrode. Because Libet was interested in recording responses associated with moving a finger, these electrodes were placed over the supplementary motor area (SMA) at the top of the brain, which controls movement. In

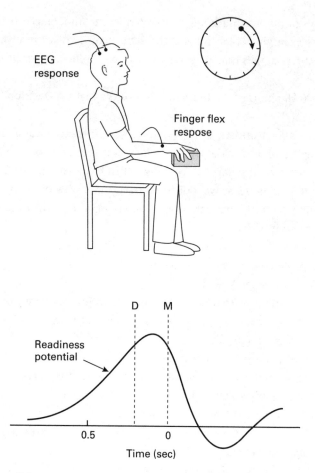

Figure 3.4

(a) Participant in Libet's experiment. (b) The electrical signal recorded from the brain. D indicates the time that the person said they had made the decision to move their finger. M is the time that the movement occurred. "Readiness potential" indicates the early part of the electrical signal that precedes D.

The Hidden Mind

addition, electrodes on the arm measured the electromyogram (EMG), which indicates flexion of arm and finger muscles.

After hearing a "get ready" tone, participants saw a moving spot on the clock that made a full revolution about every three seconds. The participant's task was simple: "Flex your finger whenever you feel like it, and note the clock position where the revolving spot was when you *first became aware of the urge* to flex your finger."

A diagram indicating the results of Libet's experiment is shown in figure 3.4b. Line D is for "decision," the time when the participants indicated they became aware of the urge to move their finger. (Libet labeled the line W for "will," but we will use D for "decision.") Line M is the time when the finger has moved. From the relation between D and M, we can see that movement occurs about 200 milliseconds (1/5 second) after the decision—the "urge" to flex the finger. This makes sense, because we would expect that the movement would occur just slightly after the decision. But what is extraordinary about the result is that the EEG signal recorded from the brain begins rising about 350 milliseconds before D. This electrical signal that occurs before the participant reported having the urge to move is called the *readiness potential*.

Libet describes this result as follows: "It is clear that neuronal processes that precede a self-initiated voluntary action, as reflected in the readiness potential, generally begin *before* the reported appearance of conscious intention to perform that specific act" (635). This is an amazing statement because Libet is saying that the brain starts responding before the person has decided to carry out the action.

While the first reaction to such a result might be to wonder if there might have been some mistake, it turns out that many other researchers have confirmed the existence of the readiness potential.[18] For example, Itzhak Fried and coworkers recorded electrical signals from single neurons in the supplementary motor area of patients who were scheduled to undergo surgery for intractable epilepsy.[19] Recording from neurons before this surgery is a routine procedure used to determine sites where seizures are being generated, to determine the regions targeted for surgical removal.

While their neurons were being monitored, Fried's patients experienced Libet's procedure. They watched a rotating clock, noted when they felt the urge to press a key, and then pressed the key. The results were similar to Libet's. The neuron's firing rate increased just before D, depending on the neuron. Fried describes his result as follows: "Several hundreds of milliseconds prior to volition, a neural process, explicit at the single neuron level, is set into motion."

What do these results mean? Libet interpreted his results to mean that brain signals occur unconsciously before people are conscious of making a decision. In other words, unconscious activity occurs in the brain just before we "decide." This conclusion seems to violate our assumption that we have some control of our decisions. Chun Siong Soon and coworkers described the conundrum posed by Libet's results as follows:

> The impression that we are able to freely choose between different possible courses of action is fundamental to our mental life. However, it has been suggested that this subjective experience of freedom is no more than an illusion and that our actions are initiated by unconscious mental processes long before we become aware of our intention to act.[20]

This is getting serious, because it involves losing our freedom! Are we, in fact, being controlled by unconscious brain processes that are determining when we will make a decision? Well, let's not be so quick to jump to this conclusion. For one thing, we have already seen that processes we are unaware of control our behavior, and later in the chapter we will see that other unconscious processes influence our behavior. Nonetheless there is something disturbing about a process that may be controlling a behavior like making a decision, which we thought we were in charge of.

As it turns out, researchers have proposed other explanations for Libet's results that do not include losing our freedom to the readiness potential. One suggestion proposes that the process of making a decision is not instantaneous but gradual.[21] We can understand the reasoning behind this suggestion by considering the process of deciding the time at which a decision has been made. Figure 3.5a shows the readiness potential.[22] The time the decision (D) and movement (M) occurred is shown at the bottom of the figure. Figure 3.5b illustrates the idea that awareness occurs in an all-or-none manner, so that D, the awareness of having made a decision, jumps abruptly from 0 (completely unaware) to 100 (completely aware). Figure 3.5c illustrates the proposal that the process is more gradual, so that the awareness of having "decided" builds up gradually, and Libet's participants say "I'm aware I've decided" when awareness reaches a threshold level, indicated by the dashed line. This idea, proposed by Jeff Miller and Wolf Schwarz, suggests that perhaps the initial part of the readiness potential indicates the initial "subthreshold" process that occurs before the decision process reaches the threshold of awareness.[23] So the conclusion is not that we aren't controlling the decision-making process; it's just that when the decision-making process is set into motion, the initial stages operate at an unconscious level.

Miller and Schwarz based their proposal on their own research and on the results of earlier experiments. Chun Siong Soon and

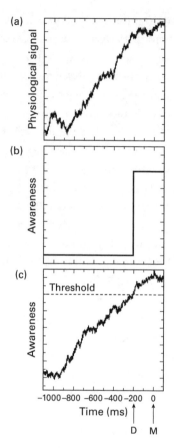

Figure 3.5

Explanation of the readiness potential proposed by Miller and Schwarz. (a) The readiness potential. (b) What occurs if awareness of the decision-making process jumps from 0 to 100 in a step function. (c) The more gradual awareness process proposed by Miller and Schwarz, in which awareness increases slowly, eventually reaching a threshold, indicated by the dashed line, which is where the person indicates that they are conscious of making a decision.

coworkers propose that "a network of high-level control areas can begin to shape a decision long before it enters awareness."[24] Patrick Haggard says that "conscious intentions come by degrees: One can be barely conscious that one is going to take the next step when walking, but intensely aware of pulling a trigger."[25]

This idea that the feeling that a decision has been made can vary from weak to strong, and can be thought of as not happening instantaneously but as building up over time until it reaches the threshold criterion, provides an alternative to Libet's conception of the relationship between the readiness potential and D.

While Miller and Schwarz's explanation seems reasonable, researchers have offered many other explanations of Libet's result as well, ranging from proposing that Libet's rotating clock does not accurately measure when the decision was made,[26] to wondering whether the readiness potential might be influenced by factors in addition to initiating a movement,[27] to musings that venture into philosophy.

One philosophical discussion involves the difference between free will (the idea that we are free to choose our courses of action) and determinism (the idea that our choices are determined by specific causes). If the brain response comes before the decision, does that mean we don't have free will? And from that, what can we conclude about whether people should be held responsible for their decisions and actions? Libet and others have thoughts about this, but as is the nature of philosophical discussions, they can become rather complicated, often with no definite conclusions.[28]

Whatever the mechanisms and implications of Libet's readiness potential, there is no question that his results have been taken seriously by researchers and have generated a great deal of research and discussion. For our purposes, we are simply going to conclude that this fascinating experiment provides yet another example of processes that are hidden from our awareness. Stated in this way, Libet's results are not that revolutionary, because as we have seen from neuropsychological research, many physiological processes operate beneath our awareness. What makes Libet's results so compelling is that they have to do with a behavior—deciding to take an action—that we feel is within our control.

We now continue our quest for hidden processes, but rather than focusing on physiology (brain damage or readiness potentials), we focus on purely behavioral experiments that reveal that there are things in the environment that we may not be aware of but can nonetheless influence our behavior.

Unconscious Processes Revealed by Measuring Behavior

Many things we are not aware of influence our behavior. One area of research on these influences considers *implicit learning*: learning that takes place without a person's awareness, which creates knowledge that is also outside the person's awareness.[29]

Implicit Learning: Learning without Knowing

Can we learn without knowing we are learning? Can we have knowledge that we are unaware of? It is easy to see that the answer to these questions is yes when we consider that you can have a grammatically correct conversation without knowing the rules of grammar. In other words, you are following rules that you learned as you were learning language, but unless you have studied grammar, you will probably find it difficult to say what these rules are.

Or consider riding a bike. If you can ride a bike, you just get on and ride. It is unlikely that you can describe the process in much more detail than "I keep my balance and pedal," but there is much more to it than that. Hidden within what is called "motor memory," your muscles know what to do. You just can't describe how they're doing it.

Returning to language, let's consider an experiment by Arthur Reber, which, although carried out when Reber was a graduate student, is recognized as a classic early experiment demonstrating implicit learning.[30] Reber's goal was to show that learning can take place without awareness, and he felt that the best way to demonstrate this was to have his participants learn an "artificial language," which consisted of letter strings created by arranging the letters *P, T, V, X,* and *S* into strings six to eight letters long (for example, *VSTPXPS*). Some of these letter strings fit grammatical "rules," which were devised by Reber and were unknown to the participants. Other strings were determined randomly. Reber found that participants presented with lists of letter strings that followed his rules made less than half as many errors when learning the lists, even though they had no idea what the grammatical rules were.

In the second part of the experiment, participants learned sequences of letter strings as before and, after learning, were told that the letter strings were created by using special grammatical rules (without telling them what the rules were). They then saw new letter strings, some of which followed the grammatical rules, and some of which were random. The task was to pick strings that were grammatical. Amazingly, participants identified the grammatical strings 70 percent of the time, even though they could not explain why; they just did it.

Just as Reber's participants were able to learn grammar without knowing what the rules were, so we learn about the characteristics of our language just by using it. Consider, for example, your ability to recognize individual words in a spoken sentence, a process called *speech segmentation*. You might think that we can tell one word from another because they are separated by silence, just as the words on this page are separated by spaces. But the fact that there are often no spaces between spoken words becomes obvious when you listen to someone speaking a foreign language. To someone who is unfamiliar with that language, the words seem to speed by in an unbroken string. However, to a speaker of that language, the words seem separated, just as the words of your native language or a language you are familiar with seem separated to you. We somehow solve the problem of speech segmentation and divide the continuous stream of the physical speech signal into a series of individual words.

One thing that helps us tell one word from another is that we know the meanings of the words. Thus, when we listen to that seemingly unbroken string of sounds of a foreign language, isolated words familiar to someone who doesn't know the language, like "gracias" or "merci" ("thank you" in Spanish and French) may "pop out" of what sounds like a meaningless string of sounds.

Word pop-out due to meaning may happen automatically, but it might not be considered implicit learning because we are conscious of words and their meanings. Other mechanisms of speech segmentation, however, are clearly implicit. One example is how people use their knowledge of the statistical regularities of language to determine where one word ends and the other begins. For example, as we learn a language, we learn that certain sounds are more likely to follow each other within a word, and other sounds are more likely to be separated by the space between two words. Consider the words *pretty baby*. In English, it is likely that *pre* and *ty* will be in the same word (***pre-tty***) and that *ty* and *ba* will be separated by a space and thus will be in two different words (pre*tty ba*by). Thus the space in the phrase *prettybaby* will most likely fall between *pretty* and *baby*.

Psychologists describe the way sounds follow one another in a language in terms of *transitional probabilities*—the chances that one sound will follow another sound. Every language has transitional probabilities for different sounds, and as we learn a language, we not only learn how to say and understand words and sentences but also learn about the transitional probabilities in that language. The process of learning about transitional probabilities and about other things that happen frequently in a particular language is called *statistical learning*.

Jennifer Saffran and coworkers carried out an early experiment that demonstrated statistical learning in eight-month-old infants.[31] During the learning phase of the experiment, the infants heard four nonsense "words" such as *bidaku, padoti, golabu,* and *tupiro*, which were combined in random order to create two minutes of continuous sound. An example of part of a string created by combining these words is *bidaku**padoti**golabu**tupiro**padoti**bidaku***. In this string, every other word is printed in boldface to help you pick out the words; however, when the infants heard these strings, all the words were pronounced with the same intonation, and there were no breaks between the words to indicate where one word ended and the next began.

The transitional probability between two syllables that appeared *within* a word was always 1.0. For example, for the word *bidaku*, when /bi/ was presented, /da/ always followed it. Similarly, when /da/ was presented, /ku/ always followed it. However, the transitional probability between the *end* of one word and the *beginning* of another was only 0.33. For example, there was a 33 percent chance that the last sound, /ku/, from *bidaku* would be followed by the first sound, /pa/, from *padoti*; a 33 percent chance that it would be followed by /tu/ from *tupiro*; and a 33 percent chance it would be followed by /go/ from *golabu*.

If Saffran's infants were sensitive to transitional probabilities, they would perceive stimuli like *bidaku* or *padoti* as words, because the three syllables in these words are linked by transitional probabilities of 1.0. In contrast, stimuli like **tibida** (the end of *padoti* plus the beginning of **bidaku**) would not be perceived as words, because the transitional probabilities were much smaller.

To determine whether the infants did, in fact, perceive stimuli like *bidaku* and *padoti* as words, the infants were tested by being presented with pairs of three-syllable stimuli. Some of the stimuli were "words" that had been presented before, such as *padoti*. These were the "whole-word" stimuli. The other stimuli were created from the end of one word and the beginning of another, such as *tibida*. These were the "part-word" stimuli.

The prediction was that the infants would choose to listen to the part-word stimuli longer than to the whole-word stimuli. This prediction was based on previous research that showed that infants tend to lose interest in stimuli that are repeated and thus become familiar, but pay more attention to novel stimuli that they have not experienced before. Thus, if the infants perceived the whole-word stimuli as words that had been repeated over and over during the two-minute learning session, they would pay less attention to these familiar stimuli than to the more novel part-word stimuli that they did not perceive as being words.

Saffran measured how long the infants listened to each sound by presenting a blinking light near the speaker where the sound was coming from. When the light attracted the infant's attention, the sound began, and it continued until the infant looked away. Thus the infants controlled how long they heard each sound by how long they looked at the light, and the infants did, as predicted, listen longer to the part-word stimuli. From results such as these, we can conclude that the ability to use transitional probabilities to segment sounds into words begins at an early age.

The idea that we follow rules we are not aware of is a topic we will return to in chapters 4 and 5, when we discuss how the "predictive mind" uses information we have learned about our environment to help us perceive the visual scenes we perceive every day. But now we consider another aspect of our everyday experience: how our behaviors—the things we do, the way we react to specific situations—are often influenced by what is going on. The situation, in which our exposure to one stimulus can affect our behavior to another stimulus, is a type of implicit learning called *priming*.

Priming: What Happens Early Affects What Happens Later

Priming is a nonconscious influence of past experience on current performance or behavior.[32] The most basic type of priming, *repetition priming*, occurs when one stimulus is presented (the priming stimulus) and then is repeated (the primed stimulus). For example, if a person reads a list of words that includes the word "something" and then is asked to complete a word starting with "some," they are more likely to answer "something" than if they had not seen the priming stimulus. Other repetition-priming experiments have shown that people respond more rapidly to words they have seen previously.

You might say that repetition priming might not be a hidden effect because a person might realize that the stimulus had been presented earlier. One way that researchers have dealt with this problem is to show that priming occurs in patients with amnesia, who, even though they have no memory of the priming stimulus, respond more quickly or accurately when that stimulus is presented again.[33]

Researchers have also reported many examples of priming where the relation between the priming stimulus and behavior is not all that obvious. For example, Aaron Kay and coworkers had one group of participants view pictures associated with business, such as desks, computers, and fountain pens, and another group view neutral pictures such as clothes, a kite, and sheet music.[34] When the participants were then asked to create a word by filling in spaces in "c _ _ p _ _ _ tive" to create a word, 71 percent of the business group created the word "competitive," compared to 42 percent of the neutral group (the other possibility, which the neutral group favored, was "cooperative").

Melissa Bateson and coworkers showed that cues of being watched enhance cooperation in a real-world setting.[35] The experiment took place in the psychology department's coffee room. Instructions on the wall at eye level indicated that tea or coffee should be paid for by dropping money into an "honesty box." This instruction had been posted for a few years before the experiment, so all Bateson had to do was add the priming stimulus. One week, photographs of two eyes were added below the instructions. The next week the pictures were changed to flowers. The eyes and flowers were alternated for ten weeks, and the amount of money deposited in the honesty box was measured at the end of each week. The results were impressive: on "eye weeks," contributions jumped up, and on "flower weeks," they fell back. The average contribution was three times greater when the eyes were present.

Based on this result, Bateson proposed that the eyes motivated cooperative behavior because they induce a perception of being watched. The eyes therefore exerted an automatic and unconscious effect on the behavior of people who worked in the psychology department. So psychologists, like other people, are susceptible to having their behavior manipulated by outside forces!

Another example of outside forces manipulating behavior is provided by an experiment by Rob Holland and coworkers, who based their experiment on the idea that some odors are associated with specific behaviors.[36] Specifically, Holland tested the hypothesis that the smell of citrus, which is associated with cleaning, will influence behaviors related to cleaning.

Holland's participants sat in a cubicle. A citrus scent of an all-purpose cleaner was diffused into the participant's space. The participant's task was to indicate as quickly as possible whether a string of letters on a computer screen was a word or a nonword. There were twenty words (e.g., *bicycling*, *hygiene*) and twenty nonwords (e.g., *poetsen*, *oprvisn*). Six of the real words were related to cleaning (e.g., *hygiene*, *cleaning*). The results of the experiment indicated that participants reacted faster to cleaning words when the citrus scent was present, whereas the scent had no effect on the noncleaning-related words.

Holland took the idea that scent can influence behavior a step farther by having participants write down five activities they were planning to do during the rest of the day. Participants who smelled the scent listed more cleaning activities (36 percent of their activities) than participants in the nonscent condition (11 percent).

This chapter began with hidden birds chirping in the trees, and now here we are at the end of a journey that has taken us through damaged brains, which reveal processes that are hidden in the intact brain; Libet's famous 1983 experiment, which is still being discussed and debated by researchers, but which most agree is yet another demonstration of "behind-the-scenes" activity of the brain; and examples that show that we can unknowingly take in knowledge, which we remain unaware of even as we are using it, and we can be influenced by environmental stimuli, also without our knowledge.

All of this is far more complicated than those hidden birds chirping in the trees (after all, those same birds do reveal themselves when they visit my backyard bird feeder). The next two chapters focus on another process that is largely hidden from our awareness: how the mind is constantly making predictions about what is going to happen next.

4 The Predictive Mind I: Perceiving and Acting

Prediction is one of the central principles of the mind's operation. It occurs in many ways, and in many different contexts. We begin by considering a situation in which a person is about to lift two boxes. The box on the left is large, and the box on the right is small, but their weights are exactly the same. The person is told to grasp the handles on top of each box and, when she hears a signal, to lift them simultaneously. What happens next is surprising, because she lifts the large box on the left much higher than the small box on the right and says that the larger box feels lighter. (Remember that both boxes are actually the same weight.) This is the *size-weight illusion*, which was first described by French physician Augustin Charpentier in 1891.[1]

Modern researchers have proposed numerous explanations for the size-weight illusion, but a recent review of the literature concluded that one of the most important causes is people's *expectations* of heaviness.[2] When observing two differently sized objects, we predict that the larger one will be heavier, so we exert more force to lift it, causing it to be lifted higher and, surprisingly, to feel lighter.

Although prediction results in an error that leads to the size-weight illusion, many of the predictions we make in everyday life are crucial to our ability to accurately perceive the world and take actions within it. In fact, the pervasiveness of prediction in everyday life has caused researchers to describe the brain as a "prediction machine," and has, especially in the last few decades, led to the proposal that prediction is an overarching principle behind cognitive abilities such as perceiving, attending, lifting and grasping, understanding language, remembering, and relating to other people.[3] In this chapter, we explore examples of the role of prediction in perception, attention, and action. In the next chapter, we consider language, music, memory, and social behaviors.

Making Predictions about the Perceptual World

> Perception is not solely determined by the input from our eyes, but it is strongly influenced by our expectations.
> —Peter Kok et al.[4]

We begin our story about prediction with the German physicist Hermann von Helmholtz (1821–1894), who was known for work in areas ranging from physics to physiology to psychology.

Helmholtz's Unconscious Inference

Helmholtz's starting point was his realization that the image on the retina is ambiguous—that is, a particular image on the retina could have been caused by a large number of different objects.[5] Consider the example of an eye looking at a book (fig. 4.1).[6] Light reflected from the book creates an image on the retina at the back of the eye, with the shape of the image determined by the geometrical process of projection: rays extended from the corners of the book meet at the lens and are extended into the eye until they intersect the retina.

The Predictive Mind I

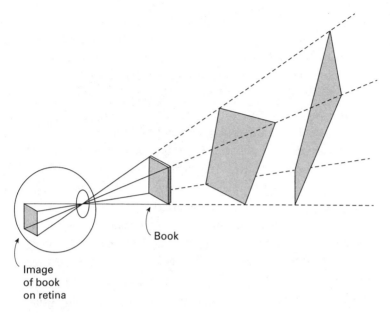

Figure 4.1
Eye looking at a book. The solid lines projecting from the book into the eye determine the book's image on the retina. The lines projecting out from the retina determine other possible objects that could cause the same image on the retina.

Although looking at a rectangular book straight on creates a rectangular image on the retina, the image becomes distorted when the book is viewed from different angles. Thus an object can create numerous images on the retina. But another, more serious problem is revealed by extending lines out from the image on the retina, as indicated by the dashed extensions of the solid lines in figure 4.1. This geometrical exercise reveals that a specific image on the retina can be caused by an infinite number of objects. Thus the rectangular image associated with the book could also be caused by the two trapezoidal objects shown or by other straight-sided objects that fit within the extended rays.

That one image on the retina can be associated with an infinite number of objects in the environment creates a problem for the brain called the *inverse projection problem*: how is it possible to determine, from the image on the retina, the object that caused the image? The solution proposed by Helmholtz is his *theory of unconscious inference*. According to this idea, we infer what object caused a particular image on the retina by applying the *likelihood principle*, which states that we perceive the object that is *most likely to have caused* a particular image on the retina.

We can appreciate how this works by considering the display in figure 4.2a. This image could have been caused by overlapping rectangles (fig. 4.2b) or by a rectangle plus an inverted L-shaped object (fig. 4.2c).[7] According to the likelihood principle, our past experiences with overlapping shapes would lead us to infer that our perception of the image in (b) is most likely caused by the two overlapping rectangles in (c).

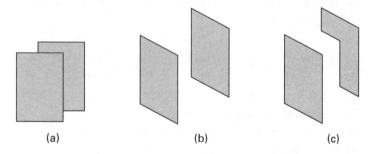

Figure 4.2
The display in (a) is usually interpreted as being (b) one rectangle in front of another rectangle. It could, however, be caused by (c) a rectangle and an appropriately positioned six-sided shape.

Helmholtz's theory of unconscious inference, which essentially proposed that perception is based on a prediction of what is most likely to be out there, was received with skepticism by his contemporaries, who rejected the idea that processes such as reasoning and inference could occur unconsciously.[8] We need to flash forward almost one hundred years to the middle of the twentieth century to observe how inference slowly made its way into the consciousness of modern-day perception researchers.

Twentieth-Century Developments

One way to appreciate how researchers thought about perception in the middle of the twentieth century is to consider physiological experiments, which focused on determining how individual neurons respond to different visual stimuli (see table 1.1c). Work by Haldan Keffer Hartline in the 1940s introduced the concept of *visual receptive fields*, where a neuron's receptive field is the area on the retina that, when stimulated, causes the neuron to fire.[9] David Hubel and Torsten Wiesel built on Hartline's work in the 1960s by showing that single neurons early in the visual system respond to small spots of light, and neurons higher in the visual cortex respond to lines oriented in a specific direction or to lines moving in a specific direction.[10] These cells were called *feature detectors*, because it was assumed that they were detecting elementary features of objects. Later research, venturing to even higher levels of the visual system, discovered neurons that responded to complex geometrical objects, faces, and places such as houses or rooms.[11]

Processing that begins with simple elements, such as spots of light or oriented lines, and progresses, building from stage to stage, to create neurons that respond to more complex stimuli like faces and houses is called *hierarchical processing*. The idea that perception can be explained by hierarchical processing that begins in the retina and proceeds stage by stage to higher areas of the cortex is called *bottom-up processing*. Although this approach dominated physiological research early in the second half of the twentieth century, another explanation was emerging, based on the idea that a complete explanation of perception requires an additional type of processing called *top-down processing*: processing that occurs when knowledge from higher levels travels "downward" to affect the upward-flowing bottom-up information.

One of the early proponents of top-down processing was the British psychologist Richard Gregory, who opened the first edition of his 1966 book *Eye and Brain* with the claim that "perception is not determined simply by the stimulus pattern—rather it is a dynamic searching for the best interpretation of the available data." In line with this idea, Gregory proposed that "a perceived object is a hypothesis."[12]

The idea of perception being a hypothesis is the theme of Gregory's book, which, although being a popular paperback written for a general audience, was slightly ahead of the zeitgeist in perception research. But even though Gregory's idea of perception as a hypothesis was a direct descendant of Helmholtz's theory of unconscious inference, it took Gregory a while to enthusiastically endorse Helmholtzian unconscious inference. Helmholtz makes a few cameo appearances in the first edition, but Gregory does not mention unconscious inference until the third edition, in 1978, when he states that "the brain is a probability computer; and our actions are based on predictions of a future situation."[13] This quote, however, was hidden at the end of the book. Helmholtz had to wait until the fifth edition, in 1997, to have his ideas placed front and center, as Gregory notes that "although active, essentially Helmholtzian, accounts of perception are now dominant, this was not so a few years ago, and they are not universally held today."[14]

Gregory's embrace of Helmholtzian unconscious inference in his 1997 edition reflects what was going on in the research literature. For example, in 1997, the *Philosophical Transactions of the Royal Society of London* published a special issue titled "Introduction to 'Knowledge-Based Vision in Man and Machine,' A Discussion Held at the Royal Society" which focused on top-down processing. Then, in 2009, another issue, "Prediction in the Brain: Using Our Past to Prepare for the Future," described how top-down processing leads to prediction.[15] Thus, by the beginning of the twenty-first century, prediction had, for many researchers, become the modern version of Helmholtz's idea of unconscious inference.[16]

An important component of the modern predictive version of Helmholtzian inference is that perception is influenced by *statistical regularities of the environment*—the fact that certain things in the environment are more likely to occur together (see chap. 3).[17] Consider, for example, the blob in figure 4.3a, which becomes an object on a table in (b), a shoe on a person bending down in (c), and a car and a person crossing the street in (d). We therefore perceive identical blobs as different objects because of our knowledge of the kinds of objects likely to be found in different types of scenes.[18]

Another effect of knowledge is illustrated in figure 4.4a, which shows indentations created by people walking in the sand. But turning this picture upside down, as in figure 4.4b, transforms the indentations into rounded mounds.[19] Our perception in these two situations can be explained by the *light-from-above assumption*: we usually assume that light is coming from above, because light in

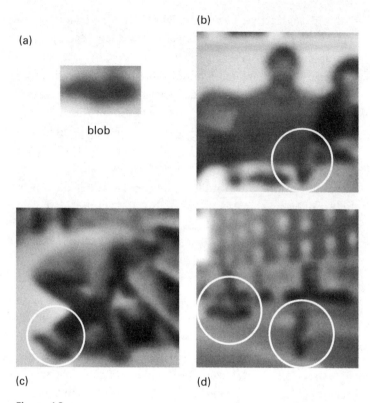

Figure 4.3
The multiple personalities of a blob: (a) the blob; (b) the blob as a bottle; (c) the blob as a shoe; (d) one blob is a car, the other, a person crossing the street. The blob takes on different identities depending on the context in which it appears.

The Predictive Mind I

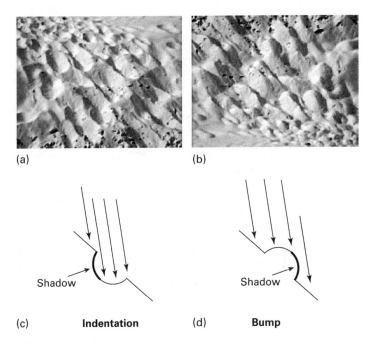

Figure 4.4
(a) Indentations made by people walking in the sand. (b) Turning the picture upside down turns indentations into rounded mounds. (c) How light from above and to the left illuminates an indentation, causing a shadow on the left. (d) The same light illuminating a bump causes a shadow to the right.

our environment, including the sun and most artificial light, usually comes from above.[20] Figures 4.4c and 4.4d show how light coming from above and from the left illuminates an indentation, creating a shadow on the left, and how the same light illuminates a bump, creating a shadow on the right. Our perception of illuminated shapes is therefore influenced by how they are shaded and by the brain's assumption that light is coming from above.[21]

The idea that the Helmholtzian process of unconscious inference is occurring every time we look around is difficult for some people to accept. Enter Michael May, a business executive who was blinded by a chemical explosion at the age of three but regained some of his vision forty-three years later after receiving a stem cell transplant followed by a corneal transplant. Just after receiving the corneal transplant, May was able to perceive only simple movements, colors, and shapes. But two years later, he said, "The difference between today and over two years ago is that I can better guess what I'm seeing. What is the same is that I'm still guessing." May was aware that he was guessing because his perceptual process was slowed down by the fuzziness of his vision, and also probably by damage to his visual system caused by more than forty years of disuse. But as L. F. Barrett and Moshe Bar point out: "What Mr. May did not know is that sighted people continually make the guesses he was forced to make with effort." Barrett and Bar point out that the brain contains representations of the environment created by experience and that "the brain uses these stored representations, almost instantaneously, to predict continuously and unintentionally what incoming visual sensations stand for in the world."[22]

But what, exactly, is the process by which we make such predictions? One answer to this question takes us back to the eighteenth-century mathematician Thomas Bayes (1701–1761). In a paper published posthumously in 1763, Bayes proposed what is known as *Bayes' theorem*, which states that our estimate of the probability of an outcome is determined by two factors: (1) the *prior probability*, or simply the *prior*, which is our initial belief about the probability of an outcome; and (2) the extent to which the available evidence is consistent with the outcome. This second factor is called the *likelihood* of the outcome.[23]

To illustrate how Bayes' theorem works, let's consider how a person we will call Mary might think about what is causing her friend Chuck's persistent cough. Mary believes that a persistent cough might be caused by a having a cold or heartburn but is unlikely to be caused by lung disease. These are Mary's priors for "causes of a cough." But looking further into possible causes, she does some research and finds that coughing is often associated with having either a cold or lung disease, but it isn't associated with heartburn. Mary considers this additional information, which is the *likelihood*, and along with her priors, concludes that Chuck probably has a cold. In practice, Bayesian inference involves a mathematical procedure in which the prior is multiplied by the likelihood to determine the probability of the outcome. Thus people start with a prior and then use additional evidence to update the prior and reach a conclusion.[24]

Although Bayes' original equations were not concerned with perception, and Bayes' ideas were ignored for at least a century and a half,[25] in the twentieth century his equations were adapted to explain behavioral inferences such as our example of Mary and Chuck, as well as examples focusing on perception.[26] For example, we can apply Bayes' idea to object perception by returning to figure 4.1. Assume that you are looking at a book on your desk. The problem is how to determine what is "out there" that is causing the ambiguous image on your retina. Luckily, one of your priors is that books are rectangular. Thus your initial belief is that it is likely that the object is rectangular.

The *likelihood* that the object is rectangular is enhanced by additional evidence such as the object's retinal image, how the image changes with changes of viewpoint and distance, and the fact that the object is on a desk. If this additional evidence is consistent with your prior that the object is rectangular, the likelihood of rectangular is high, and the perception "rectangular" is strengthened. Note that you are not necessarily conscious of taking this information into account; it occurs automatically and rapidly. The important point is that while the retinal image is still the starting point for perceiving the shape of the book, adding your prior beliefs reduces the number of possible shapes that could be causing that image.

What Bayesian inference does is to restate Helmholtz's idea—that we perceive what is most likely to have created the stimulation we have received—in terms of probabilities.[27] More specifically, Bayes specifies how to combine new with old information. We start with a hypothesis, and Bayes tells us how to reevaluate the probability of the hypothesis in light of additional evidence.[28]

I have described how inference is used to predict what is out there in the world. But how exactly is this inference implemented by neurons in the brain?

Neural Aspects of Prediction

What is the neural story behind prediction? Attempts to answer this question remind me of the idea, which I first heard from one of my professors in graduate school, that the more treatments that have been proposed for a particular disease, the less we understand how to cure it. This idea applies to discovering the neural basis of prediction, because while there are numerous ideas about what might be happening, there is little agreement about exactly what does happen.[29] I will describe a few of the ideas that have been proposed, keeping in mind that the study of the neural basis of prediction is still in its infancy.

A logical starting place for thinking about the neural basis of prediction is the distinction between top-down and bottom-up processing. Behavioral experiments have inspired a number of proposals for how higher-level neural processes might inform upward-traveling neural responses. One such proposal is based on Moshe Bar's finding that when people were asked to recognize rapidly flashed pictures of familiar objects, signals sent from the visual cortex arrived at the prefrontal cortex, which is associated with placing objects in categories ("that's an umbrella-like shape"), 130 milliseconds after the picture was flashed, whereas signals traveling along a slower pathway did not arrive at the temporal cortex, which is associated with object recognition ("that's a parasol"), until 180 milliseconds after the flash.[30]

Bar suggests that when the signal from the visual cortex arrives at the temporal lobe, it is met by the higher-level signals, which have already arrived from the prefrontal cortex (see fig. 2.1), and the information provided by these higher-level signals helps the temporal lobe determine that the object is a particular type of umbrella (in this case, a parasol). Bar calls the process by which the frontal cortex adds top-down information to the bottom-up information arriving from the visual cortex *prefrontal modulation.*

But just saying that higher-level information influences interpretation of lower-level signals leaves unanswered the question of how this interaction between higher- and lower-level information takes place. One approach to this problem is called *predictive coding.* The starting point for predictive coding is the idea that the brain contains models of the world that reflect statistical regularities of the environment. Predictive coding proposes that signals flowing upward from the receptors are compared to the model's information that is flowing downward from the brain, and the difference between the upward-flowing signals and the downward-flowing model is the *prediction error* (fig. 4.5a). This prediction error is then sent up to higher-level areas of the visual system to inform the higher-level model of changes needed to

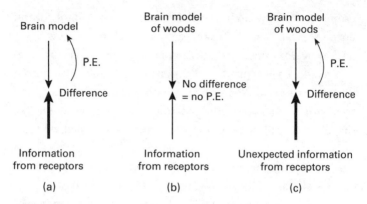

Figure 4.5
(a) Information from the receptors, flowing up, meets information representing the brain's model, flowing down. The difference between them, which is the prediction error (PE), is sent up toward the brain to provide information to make corrections to the model. (b) In the normal woodland scene, there is no difference between the signal from the receptors and the model, so there is no prediction error. (c) When something unexpected happens, the receptor information and the model do not match, so there is a prediction error.

make the model match the incoming information. When this occurs, the model changes to provide a "best guess" of what is out there, and we perceive!

For example, imagine you are hiking in the woods. As you encounter the normal woodland scene, the incoming signals that occur fit your brain's model of what to expect in the woods (fig. 4.5b). You encounter few surprises, so little or no prediction error occurs. But when something unexpected happens, such as a loud noise that sounds like something crashing through the woods, the new incoming signal does not match the model, so it generates a prediction error (fig. 4.5c). This prediction error is sent up to higher levels to help revise the brain's model to deal with this new situation.

Prediction error therefore provides an important mechanism for learning. We encounter an unexpected situation, and the prediction error provides information that enables us to make the necessary corrections to deal with the new situation. Scientists have studied neural mechanisms for prediction error physiologically by recording neural activity in situations involving prediction. For example, "prediction neurons" have been discovered that change their firing rate when monkeys receive an unexpected reward like a drop of juice or, conversely, when an expected drop of juice does not happen.[31] Researchers have also found neurons that respond to unexpected perceptual images. Travis Meyer and Carl Olson trained monkeys on pairs of images so that seeing one image (e.g., a mushroom) predicted the next image (e.g., a butterfly).[32] After this training, the butterfly caused a small response when it followed the mushroom but a larger response when it surprisingly followed a picture that the monkey had not previously seen. Results such as this have led researchers to conclude that unpredictability—or, in predictive terminology, a prediction error—is signaled by larger responses.[33]

Note the adaptive nature of the larger response to a prediction error. It is important to be able to detect the unexpected, because it is more likely to be dangerous. So when the peacefulness of our hike in the forest is suddenly interrupted by a loud noise, we need to be able to determine what is happening and react quickly, in case the noise is signaling a dangerous animal. Also, if we have an accurate idea about what is out there or what is about to happen, as we do in the normal, quiet forest environment, we can think of the visual system as being on "automatic," which requires fewer processing resources. Applying this idea to highway driving, the brain may devote fewer resources while driving on a familiar highway, thus saving processing capacity for unexpected events, like a deer crossing the road.[34]

I have been describing how information received from the environment flowing up from lower to higher levels in the visual system is compared to downward-flowing expectations based on our model of the world. But current models of predictive processing are more complex than the statement "prediction error equals the difference between information coming in and our model of the world." Some researchers have proposed that this comparison of inward-flowing information to downward-flowing information occurs not just at one place, as shown in figure 4.5, but at each level of the visual system. This process, which is called *hierarchical predictive processing*, means that prediction errors are calculated at many places along the pathway from the receptors to the brain.[35]

Because prediction error is calculated at different levels of the system, different types of information are involved at each level. For example, at lower levels of the visual system, neurons respond to simple features of a visual stimulus, such as its length or orientation. However, at higher levels, neurons respond to specific types of stimuli, such as faces, tools, or places. Needless to say, hierarchical predictive processing multiplies the complexity of neural calculations that occur, and a number of complicated equations and neural-flow diagrams have been proposed to explain how predictive processing works in vision.[36] But even though these diagrams are based on research that provides evidence for prediction error, the details of exactly how this error information changes our representations are still speculative, and descriptions of how neurons accomplish this are still sketchy.[37]

Let's stop and consider how thinking about visual processing has changed since the 1950s. Initially, the idea of a hierarchical approach to visual processing, which involved increasingly complex feature detectors at higher levels of the visual system, dominated thinking. This approach saw processing as largely bottom-up, with one level building on the other as signals moved to higher and higher levels of the visual system. But the new emphasis on top-down processing, and the mechanisms proposed by predictive coding, have upended this linear one-way picture of visual processing. Present-day thinking sees higher-level information as being crucial, with downward-flowing signals playing a crucial role in the visual system.[38] As Andy Clark, a proponent of predictive coding, states, "Brains are not fundamentally in the business of processing inputs—they are in the business of predicting their inputs."[39]

Moving beyond Stationary Perception

We started our journey of prediction with Helmholtz's insight that the ambiguity of the retinal image can be dealt with through inference. While ideas involving concepts such as prediction error and hierarchical predictive processing have put a modern face on Helmholtz's ideas, the most persuasive evidence for the role of prediction in perception (and, as we will see later, cognitive processes in general) is provided by how prediction can explain our interactions with the environment.

We can interact with the environment with our mind, as when we observe something happening and think about what it means. But most interaction involves engaging physically with the environment, and this engagement involves movement, which often depends on prediction. Our starting point is the eyes, which are almost constantly in motion.

The Moving Eyes

In many situations, we need to move our eyes as we interact with the environment. For example, when we look at a scene, we typically scan the scene to identify objects and to determine their relations to one another.

Consider a record of the path of a person's eye movements as the person scans a scene (fig. 4.6).[40] Each dot represents a *fixation*, during which the eye is at rest for a fraction of a second. Each line represents a *saccade*, during which the eye is rapidly moving from one fixation to the next. People typically make two to three fixations per second when exploring a scene.

The reason we are almost constantly moving our eyes is that there is one place on the retina, the fovea, where the image of an object falls when we look directly at it. Because the fovea is

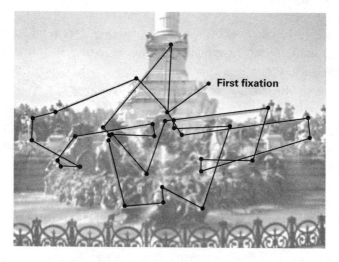

Figure 4.6
Scan path of an observer freely viewing a picture. Fixations are indicated by dots, and saccadic eye movements by lines.

responsible for high-acuity vision, we move our eyes to direct the fovea at whatever we want to see clearly. Because eye movements indicate shifts of attention, they provide a window into our mind, because it is the mind that is controlling our eye movements, and these eye movements provide evidence for the predictive nature of attention.

Initial studies of visual attention showed that *salience*—physical properties that make particular parts of a scene stand out—is a powerful determinant of attention. For example, a bright patch of sunlight illuminating a dark corner of the forest, an animal scampering from one place to another, or a color that stands out from a homogeneous gray background—all these things initially capture our attention. But the power of salience fades with time, as higher-level factors, based largely on acquired knowledge and prediction, take over. Here are a few examples:

Searching a familiar scene. Searching for a particular object in a scene is often guided by knowledge of where things usually appear in a scene. For example, when searching for a cup in a kitchen, we concentrate our eye movements along the counter; when looking for paintings in a room, our gaze moves across the wall.[41] In other words, fixations are controlled by statistical regularities of the environment.[42]

Encountering unpredicted objects. Some things stand out because they aren't supposed to be where they are. When confronted with a kitchen scene in which a computer printer is sitting on top of a stove, our attention is drawn to the printer. A pot on the stove, however, does not elicit special attention. Things that violate our expectations attract attention and, as we saw earlier, are also associated with prediction error.[43]

Task-related attention. Measuring eye movements during a task reveals how our attention shifts as the task progresses. For example, the process of making a peanut butter and jelly sandwich begins as we retrieve two slices from a loaf of bread and place them on a plate. Our eyes move to follow this action and then shift to the peanut butter jar just before we lift it, and to the jar's lid just before we remove it. Thus we direct eye movements at parts of the scene relevant to the task, and these eye movements usually precede the action by a fraction of a second—an example of the "just in time" strategy, based on predictions of what is about to happen.[44]

The story of eye movements and attention illustrates how movements of the eyes are based on both characteristics of the scene and prediction based on our knowledge of the world. But while movements of the eyes provide us with the ability to scan a scene, the sequences of fixations and saccades that direct the mind's attention to different places in a scene create a problem for the mind to solve.

The problem occurs when the eyes move from one fixation to the next. Imagine that you are looking at a flower in a garden and then move your eyes to fixate on a rock about four feet away. Looking at the flower creates an image of the flower on your fovea. Moving your eyes to the rock causes the rock to replace the flower on your fovea. The problem occurs between these two fixations, because as the eyes move from flower to rock, the objects located between the two sweep across the fovea. Nonetheless, we perceive no movement. The scene remains stationary.

The reason we don't perceive the scene as moving was explained by Helmholtz as follows: The eyes move because a signal to move is sent from the motor area in the brain to the eye muscles (fig. 4.7). Simultaneously, a similar signal is sent to the area of the brain responsible for perception. This signal, which Helmholtz called the *efference copy*, but which is more commonly called the

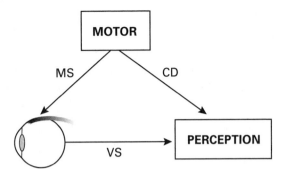

Figure 4.7
The basic principle behind why we don't see a smeared image when we move our eyes from one place to another. To make the eyes move, a motor signal, MS, is sent from the motor area to the eye muscles. Simultaneously, a copy of this signal, the corollary discharge, CD, is sent to the brain area responsible for perception. When the eyes move, a visual signal, VS, indicating movement of the scene across the retina, is sent to the area responsible for perception. However, the corollary discharge meets this signal and inhibits perception of the smeared image caused by movement of the eye. Note that this is a highly simplified diagram, as there are multiple pathways and areas involved in both the motor and perceptual aspects of this process.

corollary discharge by modern researchers, reaches the perception area just before the eye moves and so predicts that the eye is about to move.[45] This prediction is essentially telling the brain, "The eye is going to move, not the scene, so the scene should remain stationary." The corollary discharge therefore inhibits perception of movement created by the movement of the eye.[46]

This mechanism, which tells the brain that movement of the image across the retina is caused by movement of the eye and not the scene, leads to the conclusion that if the eye were to move without any accompanying corollary discharge, the scene should appear to move. You can accomplish this by closing one eye and gently pushing the eyelid of the open eye, so the eye moves slightly. When you do this, the eye moves with each push, but since there is no signal to the eye muscles or corollary discharge, you perceive the scene as moving. Imagine how disturbing it would be if this kind of movement occurred each time your eyes moved while scanning a scene. Prediction provided by the corollary discharge therefore helps create a stable world even though our eyes are moving.[47]

It is clear that rapid prediction is an essential mechanism for humans, and it also occurs in many other species as well. One example of the corollary discharge at work in another species is provided by the cricket. Male song crickets create songs to attract females and warn off rival males by rubbing their forewings together. Their call is a chirp consisting of three to five pulses that last about 250 milliseconds (1/4 second) each, followed by 300 to 500 milliseconds of silence before the next chirp. One feature of the chirp is that it is intense—about 100 decibels.[48]

The cricket's ears are on its legs, near the source of the chirp, so it would seem that the cricket's auditory system could potentially be deafened by its own chirp. However, a corollary discharge signal comes to the rescue by suppressing transmission in the auditory pathway during each chirp. Thus, just as a corollary discharge suppresses the perception of movement when a person's eyes move, so a corollary discharge suppresses perception of sound when a cricket chirps. This suppression occurs only during the chirp, so the cricket can hear during the silent interval between chirps.

Corollary discharge signals clearly are important components of actions created when the eyes move, but predictive mechanisms are also involved in other interactions with the environment. They also affect how we experience sensations such as touching and being touched.

Touching and Being Touched

When someone touches you, their touch feels more intense than when you touch yourself with the same force.[49] You can understand why this occurs by considering what happens when you position your finger about an inch above your arm and move your finger down, so that it taps your arm. The movement of your finger toward your arm occurs because a command is sent from your motor cortex to your finger muscles, and at the same time a corollary discharge containing information about the downward motion of your finger is sent to brain areas such as the somatosensory cortex that are responsible for feeling touch. What you feel is therefore determined by both sensory stimulation of the skin and the corollary discharge warning signal, which decreases the sensation of touch. If, however, someone else touches your arm, your tactile sensation is determined solely by sensory stimulation. The process is similar to what happens with saccades. Eye movements accompanied by a corollary discharge result in no perception of visual smear; tactile sensations accompanied by a corollary discharge result in a lesser perception of tactile pressure and also less activity in the somatosensory cortex.[50] The fact that other-touching feels stronger than self-touching has an adaptive function, because being touched by someone else is potentially more dangerous than touching yourself.

The difference between self- and other-touching has another consequence: it is difficult to tickle yourself! You can demonstrate this by first gently moving your fingers over your arm while observing the motion and noting the resulting sensation. Then close your eyes and ask another person to do the same thing to your arm. You might notice that when the other person applies the stimulation, the tactile feeling is different and perhaps more ticklish.

The difference between self- and other-tickling was demonstrated by Sarah-Jayne Blakemore and coworkers, who used a device that presented gentle stimulation to a person's palm.[51] When the person controlled the stimulation, he or she felt tactile sensation but little "tickle." However, when the device controlled the stimulation so that it was unpredictable, the tactile sensation felt more ticklish. The explanation, in terms of the presence of corollary discharge (for self-tickling) and the absence of corollary discharge (for other-tickling), is the same as the explanation for touching. What is particularly interesting about the tickle experiment is that other-tickling changes not only the *magnitude* of tactile sensation compared to self-tickling but also the *quality* of the sensation.

Acting on the Environment

Moving your eyes to scan a scene and experiencing touch are important aspects of our experience. But to fully experience the world, we have to act within it and on it. We walk from one place to another, pick things up, and manipulate objects. To illustrate how prediction works in situations like these, we will focus on the scenario pictured in figure 4.8, in which a person (a) reaches for a ketchup bottle with her right hand, (b) grips the bottle, and then (c) lifts and positions it, so that (d) the left hand can hit it to deposit a dollop of ketchup onto a hamburger.[52] To understand the role of prediction in this process, we will consider each component of the process separately. We begin with reaching for the bottle.

When you decide to reach for something, be it a bottle of ketchup or a cup of coffee, the process appears simple on the

The Predictive Mind I

Figure 4.8
Steps leading to depositing a dollop of ketchup on a hamburger. The person (a) reaches for the jar with her right hand, (b) grasps the bottle, (c) lifts the bottle, and (d) positions the bottle over the hamburger and delivers a "hit" with her left hand to dispense the ketchup. (Photo credit: Bruce Goldstein.)

surface. You move your hand toward the object and grasp it. But hidden behind the simplicity of this everyday occurrence is a sequence of processes that are far from simple. As a person decides to reach for a bottle of ketchup, a signal is sent from an area of the brain called the *parietal reach region* (PRR), which contains neurons that indicate where the person is about to reach.[53] Because this occurs before the reach happens, it can be described as predictive of the coming reach. Once the reach begins, prediction continues as the visual system monitors the hand's location, so if the reach veers off course, causing a prediction error, appropriate course corrections can be made.

A similar predictive mechanism operates for grasp. Neurons called *visuomotor grasp neurons* fire to seeing a specific shape and also fire as the grip is adjusted to match the object's shape.[54] Thus the shape of the grasp is determined before the hand makes contact with the bottle.

Once the bottle is grasped, it is time to lift it, and the force of the lift is determined by a predictive process that takes into account the size of the bottle, how full it is, and past lifting experiences with similar objects. Thus different predictions occur if the bottle is full or almost empty, and if the prediction is accurate, then the bottle will be lifted with just the right force. However, if the prediction is not accurate, which would occur if the person thinks the bottle is full when it is almost empty, then a prediction error occurs, and the lift will be too forceful and therefore too high (as in the size-weight illusion that opened this chapter), so a correction needs to be made.

Nearing the end of this sequence, the bottle is now positioned above the hamburger, but because the ketchup refuses to dispense, it is encouraged by a swift hit delivered by the left hand (fig. 4.8d). Accurate prediction of the force of the blow helps ensure that the right hand holds the bottle firmly in place when it is hit. This information is provided by the corollary discharge that accompanies the signal sent to the muscles of the left hand.

What happens if someone else hits the ketchup bottle? If this happens, there is no corollary discharge to signal the strength of the hit, so the holder's grip may or may not be strong enough to hold the bottle firmly, creating the possibility that not only the ketchup but the bottle itself might be propelled toward the hamburger.[55]

Simple actions that we carry out every day therefore depend on constant interactions between sensory and motor components of the nervous system, and constant prediction of what is going to happen next: how far to reach, how to adjust your hand to grasp things properly, how hard to grip, all of which depend on the objects involved and the upcoming task. Thus, picking up a pen to write involves a different configuration of the fingers and force of grip than picking up the ketchup bottle or picking up a pen to move it to another place on your desk. Considering the number of actions you carry out every day, there is a lot of prediction going on. Luckily you usually don't have to think about it, because your brain takes care of it for you. In the next chapter, we carry our discussion of prediction to more cognitive activities such as language, memory, and social interaction.

5 The Predictive Mind II: Language, Music, Memory, and Social Prediction

The predictive nature of the mind extends beyond influencing where we look, keeping our world stable and enabling us to move through the environment and manipulate objects within it. This chapter extends the scope of prediction into the higher cognitive realms of language, music, memory, and social interactions.

We first consider language: what happens when you ask the person across the table, "Is that enough ketchup?" You might elicit an answer and possibly begin a conversation. Or perhaps you are alone and decide to read as you eat. In either case, you have embarked on a journey of language, which provides many examples of prediction. After language, we will consider music, which, like language, consists of a string of sounds. We will then look at how prediction plays an unexpected role in remembering the past and perhaps a less unexpected role in guessing the future. We will then consider the crucial role of prediction in the social realm, which involves interacting with other people.

Language

Language involves multiple scales of meaning, ranging from sounds that create meaningful words, to strings of words that create sentences, to multiple sentences, which can create stories. To appreciate this idea, consider the statement that begins, "My new puppy Nala is chew..." If you added "ing," you will have made a correct prediction, based on the rules of grammar. When the sentence continues, "chewing on...," your next prediction is less constrained, as there are many ways to continue, such as "my foot," "my furniture," or "my penguin," all of which are grammatically correct, but one of which is much less likely than the others (although the situation would be different if the lead-up to the sentence had placed the locale as Antarctica, mentioned your pet penguin, and was changed to begin "My new puppy Nala is chas..."). These examples illustrate how we can make linguistic predictions and how they depend on the initial words of the sentence and the context in which they appear. Language, whether spoken or read, involves a sequence of one prediction after another. Let's consider a few examples.

Perceiving Strings of Words

At the end of chapter 3, I introduced the idea that our knowledge of the transitional probabilities between the sounds of a language enables us to segment a string of spoken sounds into individual words, a process called speech segmentation.

Another predictive mechanism for speech segmentation is the context of a sentence. For example, consider the following two sentences:

(1) Sally is a big girl.
(2) Big Earl loves his car.

Although the speech signal for the last two words in (1) and the first two words in (2) can be identical, the sounds are separated differently in our minds. Another example is "I scream, you scream, we all scream for ice cream," in which "I scream" and "ice cream" are segmented differently based on the sentence's meaning.

Predicting Upcoming Words

Another way that context can influence language is illustrated by our example of the sentences involving Nala, the dog, in which predictions were informed both by the beginning of the sentence and by the context within which the sentence appeared. One method of measuring these effects is determining *cloze probability*, the probability that an upcoming word will be guessed based on the first several words of a sentence.[1] In an experiment by Keith Rayner and coworkers,[2] cloze probabilities were determined by presenting a partial sentence and asking a group of participants to guess the next word. A word's cloze probability is the percentage of participants who pick that word. For example, consider the following two pairs of sentences, in which participants were asked to supply the target word following "the" in the second sentence:[3]

Predictable condition:
Hank is scared of eight-legged bugs.
He screamed when he saw the ———.

Unpredictable condition:
Hank has always been a fearful person.
He screamed when he saw the ———.

In the predictable condition (where *spider* is the predicted word), the cloze probability for spider was 0.70, whereas in the unpredictable condition, the cloze probability for spider was 0.125. In the second part of the experiment, when participants' eye movements were measured as they read these sentences, the location of the target word was more likely to be skipped or receive shorter fixations in the predictable condition compared to the unpredictable condition. Thus, as people read, and also when they engage in conversation, they are continually making predictions about what words are coming next, with some words being favored over others.

Prediction can also be revealed by *where* we look. For example, when participants heard "The boy will eat the cake" while viewing a picture like the one in figure 5.1, their eyes fixated on

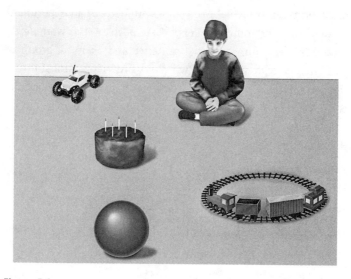

Figure 5.1
A picture similar to one used in an experiment in which eye movements were measured as participants heard a sentence while looking at the picture. Hearing the word "eat" in the sentence "The boy will eat the cake" generates a rapid prediction so the listener's eyes arrive at the cake before hearing the word "cake." This prediction does not occur in response to "move" in the sentence "The boy will move the cake."

the cake 87 milliseconds before hearing the word *cake*. In contrast, when hearing "The boy will move the cake," fixation on the cake was delayed until 127 milliseconds *after* hearing *cake*. Thus, hearing the word *eat* causes participants to zoom in on the only edible object in the scene before hearing the next words.[4]

The predictability of upcoming words is also signaled by electrical responses in the brain. The *event-related potential* (ERP), which is recorded with disc electrodes on a person's scalp, has a number of components. One of these components is called the N400 wave because it is a negative response that occurs about 400 milliseconds after a word is heard or read. One of the characteristics of the N400 response is that it is larger to an unexpected word.

The type of result that occurs when measuring a person's N400 while they listen to two type of sentences is shown in figure 5.2[5] The solid line shows that there is little or no response to *a* in the sentence "There was a nice breeze so the girl went out to fly a kite." The dashed line shows the large N400 response to *an* in the sentence "There was a nice breeze so the girl went out to fly an airplane." The reason for the difference is that "a kite" has a high cloze probability (it is the most likely ending to the sentence), whereas "an airplane" has a low cloze probability. The brain therefore anticipates the most probable continuations for sentences in advance of the actual input. Another way to state this is that the representations of the most likely words are pre-activated in advance of their appearance.[6]

Understanding Sentences

In addition to anticipating words in sentences, readers or listeners anticipate the *structure* of a sentence—that is, how the words in a sentence are organized into phrases in a person's mind, a

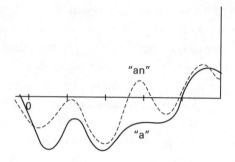

"There was a nice breeze, so the girl went outside to fly..."

Figure 5.2
N400 ERP resonse to the words *a* (solid line) and *an* (dashed line) when added to the end of the sentence "There was a nice breeze so the girl went outside to fly..." The word *an*, which has a low cloze probability, generates a larger ERP, which can be conceptualized as an "error response."

process called *parsing*.[7] What does parsing tell us about prediction? Consider a sentence that begins as follows:

(1) After the woman played the piano...

What comes next? If you expect a phrase like one of these,

...she got up and took a bow.

...she left the stage.

...the crowd cheered wildly.

then you would be organizing the sentence into two phrases, like this:

(2) [After the woman played the piano] [she left the stage]

However, what if the sentence in (1) continued

...was wheeled off the stage.

You probably didn't expect that ending, which creates the following two phrases:

(3) [After the woman played] [the piano was wheeled off the stage]

Notice that the first phrase in (3) ends with *played*, whereas our first guess at parsing, shown in (2), included *the piano* in the first phrase.

The sentence "After the woman played the piano was wheeled off the stage" is an example of a *garden path sentence*, because the reader is "led down the garden path" (a poetic way of saying "misled") and so initially parses the sentence incorrectly. Garden path sentences illustrate prediction gone wrong. Because construction (2) occurs more often for a sentence that begins as in (1), we predict that is what will occur, only to be taken by surprise when it does not.

Why does garden pathing occur? The answer to this question is complex, because a number of different situations make garden pathing more likely.[8] One thing that affects garden pathing is the meaning of words in a sentence. Consider these two sentences:

(1) The defendant examined by the lawyer was unclear.
(2) The evidence examined by the lawyer was unclear.

Sentence 1 is more likely to lead you down the garden path because two possibilities present themselves after reading "The defendant examined." The defendant could be being examined by someone else, as in our example, or the defendant could be examining something, as in "The defendant examined the courtroom." Only after reading the rest of the sentence, "...by the lawyer," is it possible to definitely determine that it is the defendant who is being examined. In contrast, only one possibility presents itself after reading "The evidence examined" in sentence 2, because it is unlikely that the evidence will be doing any examining.

The likelihood of garden pathing is also influenced by the context in which a sentence appears. Consider, for example, the following sentence proposed by Thomas Bever (1970), which has been called the most famous garden path sentence, because of the confusion it causes:[9]

The horse raced past the barn fell

Whoa! What's going on here? For many people, everything is fine until they hit *fell*. Readers are often confused and may even accuse the sentence of being ungrammatical. But let's look at the sentence in the context of the following story:

Two jockeys decided to race their horses. One raced his horse along the path that went past the garden. The other raced his horse along the path that went past the barn. The horse raced past the barn fell.

Of course, the confusion could have been avoided by simply stating that the horse *that was* raced past the barn fell, but even without these helpful words, context wins the day, and we parse the sentence correctly.

Because prediction is not 100 percent accurate and can therefore sometimes lead to errors, you might wonder if prediction is a good thing. The answer is that we are constantly making predictions about what is probably going to come next in a sentence, and because our predictions are based on what is most likely to happen, most of the time we are right. This is basically the idea behind Helmholtz's likelihood principle, which proposed that we perceive what is most likely. In language, as in perception, we may occasionally be fooled, but we are most often correct.

Music

> When listening to music we constantly generate plausible hypotheses about what could happen next.
> —Stefan Koelsch et al.[10]

Next time you listen to music, see if you can predict what is coming next. Even if you are hearing a piece for the first time, it's often possible to do this with surprising accuracy. People are able to make two kinds of predictions: (1) *when* notes are going to happen next, and (2) *which* notes are coming next.

Predicting Timing

The *when* of appreciating music involves perception of a regular beat, which often elicits the physical response of swaying or tapping in listeners. Regular intervals between notes are described as *meter*, which is typically indicated by time signatures like 4/4, 2/4, and 3/4. Meter provides listeners with an expectancy that guides their prediction of when the next note is going to happen.[11] From this perspective, we can consider "taking up the beat" as an act of prediction, because each tap of the foot is followed by a prediction of when the next tap will happen.

This prediction of timing is accompanied by an oscillation of brain waves, which is synchronized with the beat, so that the oscillation peaks on the beat, decreases, and then rebounds to predict the next beat (fig. 5.3).[12] Underlying this response are neurons that are tuned to specific time intervals.[13]

Not only does the brain's timing response demonstrate prediction, but it also responds to prediction error, which occurs when the expected timing established by meter is modified by syncopation, which occurs when there is a mismatch between

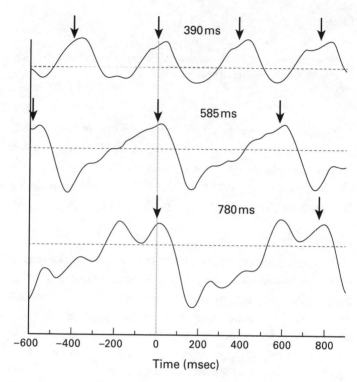

Figure 5.3

Time course of brain activity measured on the surface of the skull in response to sequences of equally spaced beats, indicated by the arrows. Numbers indicate spacing, in milliseconds, between each beat. The fastest tempo (about 152 beats per minute) is at the top, and the slowest tempo (about 77 beats per minute) is on the bottom. The brain oscillations match the beats, peaking just after the beat, decreasing, and then rebounding to predict the next beat.

the regular beat specified by the meter and the actual notes.[14] The two compositions in figure 5.4 illustrate this process.[15] Figure 5.4a shows a composition in a regular 4/4 meter. Figure 5.4b shows the same composition, but with the four quarter notes each represented by two joined eighth notes. One way to "count this out" is illustrated below the musical staff, with the arrows indicating the beat. The match between meter and notes is obvious, with the beat falling right at the beginning of each quarter note, as we count 1-and, 2-and, 3-and, 4-and. Figure 5.4c shows how syncopation, created by adding an eighth note at the beginning, changes the relation between the beat and the notes, so that the three quarter notes begin "off the beat" on the "and" count.

Syncopated rhythms like this are at the heart of jazz and pop music and have been linked to people's urge to dance, to "be in the groove."[16] Figure 5.4d shows that the brain's response to the less predictable syncopated series is larger than to a more predictable nonsyncopated series.[17] Responses to prediction and prediction error therefore occur not only in vision and language but in musical timing as well. And we will now see that responses to prediction error also extend to sequences of notes.

Predicting Notes

We now consider the *what* of music prediction: what note is going to come next?[18] Just as sequences of words are organized into phrases and are governed by syntactical rules for arranging these components, so sequences of musical notes are also organized into phrases and governed by musical syntax.[19] But although music and language both unfold over time and have syntax, the rules for combining notes and words are very different. Notes are combined based on their sound, with some sounds going together better than others, whereas words are combined

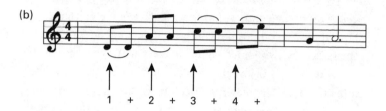
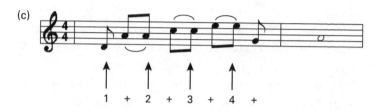
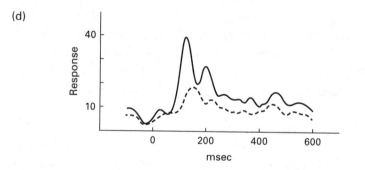

Figure 5.4

Syncopation explained. (a) The top record shows a simple melody consisting of four quarter notes in the first measure. The middle record (b) shows the same melody, with each quarter note changed to two joined eighth notes. The count below this record indicates that each quarter note begins on the beat. This passage is therefore not syncopated. (c) Syncopation is created by adding an eighth note at the beginning. The count indicates that the three quarter notes start off the beat (on "and"). This is an example of syncopation. (d) Brain response to non-syncopated melody (dashed line) and syncopated melody (solid line).

based on their meanings. No analogues exist for nouns and verbs in music, and there is no "who did what to whom" in music.[20] However, one thing that spoken language and music have in common is that listening to speech and listening to music both involve making predictions about what might be coming next as a sentence or musical composition unfolds.

One way to illustrate expectation in music is to consider how the notes of a melody are organized around the note associated with the composition's key, which is called the *tonic*.[21] For example, C is the tonic for the key of C and its associated scale: C, D, E, F, G, A, B, C. Organizing pitches around a tonic creates a framework within which a listener generates expectations about what might come next.

One common expectation is that a song that begins with the tonic will end on the tonic. This effect, which is called *return to the tonic*, is illustrated in the musical phrase with the words "Twinkle, twinkle, little star, how I wonder what you are," which begins and ends on the same note. The power of musical expectation becomes apparent when singing a song and stopping just before the song has returned to the tonic. The effect of pausing just before the end of the phrase (at *you* in *Twinkle, twinkle...*), which could be called a violation of musical syntax, is unsettling and leaves us longing for the final note for *are* to bring us back to where we started.

Another violation of musical syntax occurs when an unlikely note or chord is inserted that does not seem to "fit" in the tonality of the melody. Aniruddh Patel and coworkers had participants listen to a musical phrase like the one in figure 5.5, which contained a target chord, indicated by the arrow above the music.[22] There were three different targets: (1) an "in-key" chord that fit the piece, shown on the musical staff; (2) a "nearby-key" chord

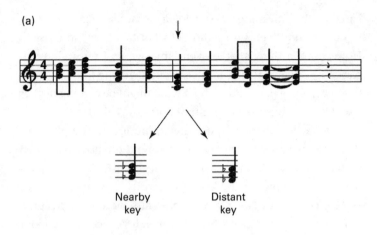

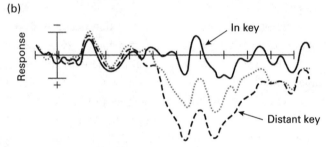

Figure 5.5
(a) The musical phrase heard by participants in Patel and coworkers' experiment. The location of the target chord is indicated by the downward-pointing arrow. The chord in the music staff is the "in-key" chord. The other two chords were inserted in that position for the "nearby-key" and "distant-key" conditions. (b) ERP responses to the target chord. *Top*, in key; *middle*, near key; *bottom*, far key. Greater deviations from the "correct" key generate larger responses. The response record shown lasts one second.

that did not fit as well; and (3) a "distant-key" chord that fit even less well. Listeners judged the phrase as acceptable 80 percent of the time when it contained the in-key chord; 49 percent when it contained the nearby-key chord; and 28 percent when it contained the distant-key chord. Listeners were apparently judging how "grammatically correct" each version was.

Patel then used the event-related potential (ERP) to determine how the brain responds to these violations of musical syntax. When we discussed the ERP in connection with the "fly a kite" sentence earlier, we saw that the N400 component of the ERP was larger in response to words that were not as likely (see fig. 5.2). Another component of the ERP, the P600, which is a positive response that occurs 600 milliseconds after the onset of a word, becomes larger in response to violations of syntax. For example, there is no P600 response to *eat* in the sentence "The cats won't eat," but there is a large response to *eating* in the grammatically incorrect sentence "The cats won't eating."

When Patel measured the ERP response to each of the three musical targets in figure 5.5, the in-key chord (top record) elicited little positive response, but the out-of-key chords elicited positive responses, with the larger response for the more out-of-key chord. Patel concluded that music, like language, has a syntax that influences how we react to it.[23]

The predictive nature of music has also been demonstrated by the cloze probability task that was originally developed to study language. Cloze probability is measured for music by having listeners sing the note they think comes next in a string of notes. An experiment using this technique showed that listeners completed a novel melody by singing the tonic note in an average of 81 percent of the trials, with this number being higher for listeners who had formal musical training.[24] Thus, as we listen to music, we are focusing on the notes we are hearing while we are creating expectations about the *where* and *what* of what is going to happen next.

Memory

> Prediction gives organisms a head start.
> —Daniel Gilbert and Timothy Wilson[25]

Memory, being the repository of all our knowledge, is essential for prediction. The most clear-cut examples are situations in which a connection between two events is learned, as when a bird associates the appearance of a cat with the tinkling of a bell. This connection later serves as a signal, so the bird can avoid the predicted cat. On a more sophisticated level, we have seen how the prediction provided by knowledge of words and situations stored in memory is an essential tool for understanding language. In these examples, memory enables prediction, as the bird or person draws from knowledge stored in memory. But beyond this basic function, prediction plays a role in *creating* memories. To understand how this works, we need to consider how memories are created.

How Memories Are Created

In a nationwide poll conducted in 2009, 63 percent of people agreed with the statement "Human memory works like a video camera, accurately recording the events we see and hear so we can review and interpret them later." In the same survey, 48 percent agreed that "once you have experienced an event, and formed a memory of it, that memory does not change."[26] As it turns out, both of these beliefs are wrong. Memory, rather than being like a photograph in which—snap!—a memory is formed, is more accurately described as a construction, which is created from information taken in when an event happens plus other information that may occur later.

The idea of "constructive memory" was demonstrated in a classic experiment, carried out before World War I, in which Frederic Bartlett had participants read a story from Canadian Indian folklore describing a battle. Bartlett's participants were asked to recall the story shortly after reading it, and then at longer intervals, ranging up to years. The key result of this experiment is that the remembered stories changed as time progressed, and the change reflected the participant's own culture, so that eventually a tale from Canadian folklore became a story straight out of the participants' Edwardian English background. Bartlett's participants were creating their memories from two sources, the story they had read and similar stories from their own culture, which led Bartlett to conclude that remembering is "an integrative reconstruction or construction."[27]

Modern experiments have confirmed the idea of "memory as construction" by showing how memories can be distorted by the knowledge that a person brings to a situation or by events that happen afterward. For example, when participants who were sitting in an office waiting for an experiment to begin were later, when in another room, asked to describe the office, 30 percent of the participants remembered seeing books, even though there were none there.[28] Results such as these are interpreted in terms of *schemas*—a person's knowledge of some aspect of the environment. Because books are part of many people's "office schema," they were "remembered" as being in the office.

In these examples, we are conceptualizing prediction as a mechanism of inference. A person who was in an office "remembers" by predicting what he or she would expect to see in an office, and uses this expectation to create a memory of the office. But memories, however created, are not etched in stone. Not only can prediction sculpt memories as they are being formed, but predictions can edit memories after they are created.

Predictive Editing of Memory
A memory, once formed, can be changed by later experiences. Consider, for example, a situation in which, driving from the airport to visit a friend, you find that your favorite route has been unexpectedly interrupted by construction, which has permanently changed the routing of your trip.

What does this mean from the point of view of prediction? Your prediction, based on memory, is that you will turn left on First Avenue. But as it turns out, First Avenue is permanently closed, so you have to find another route. Your prediction, therefore, was in error. If we recall our discussion of prediction error in chapter 4, this error provides information that can be used to form a new, updated memory, which you will use the next time you visit your friend.

Evidence for memory being modified by prediction error is provided by an experiment in which participants were exposed to three images, which could be faces or natural scenes, in sequence A, B, and C. This exposure set up the expectation that when A and B are presented, C will follow. If, at this point, the participants' memory is tested, they remember A, B, and C with high confidence. But sometimes, after training on the A, B, C sequence, participants see A and B but have their prediction that C will appear dashed when a new picture, D, appears instead. This creates a type of prediction error called *context-based prediction error*,[29] and when memory is tested, participants are less confident than before in their memory of C.

The message from this result is that a memory trace can be weakened when a prediction is not fulfilled. When C is no longer predicted, memory for C is weakened. This makes sense from an adaptive viewpoint, because if C has become less likely, weakening memory for C frees up space in the mind for forming new memories.[30]

Simulating the Future

Organisms remember the past so they can predict the future.
—Daniel Gilbert and Timothy Wilson[31]

Another link between memory and prediction has been demonstrated in experiments by Stanley Schacter and Rose Addis, in which participants were asked to imagine future events. For example, they were shown a picture of an event like a party or someone working in an office and were then asked to imagine events that could happen in the next few years, with the picture as the setting. In doing this, participants were not strictly predicting the future, but they were simulating what might happen.[32]

What, you might wonder, does simulation of the future have to do with memory? The answer is that there is a close link between the two. When Addis used fMRI to determine how the brain is activated while participants silently thought about events from the past or imagined what might happen in the future, the same brain regions were activated, suggesting that similar neural mechanisms are involved in remembering the past and imagining the future.[33] Based on these results, Schacter and Addis proposed the *constructive episodic simulation hypothesis*, which states that episodic memories—memories for a person's past experiences—are extracted and recombined to construct simulations of future events.

Supporting the idea that past memories create future simulations is the finding that patients who have suffered brain damage that caused them to lose their ability to remember events from their past find it difficult to imagine what might happen in the future. This inability to imagine future events is often restricted to things that might happen to the person personally. Imagining other future events, such as what might happen in politics or other current events, is not affected.[34]

The link between remembering the past and thinking about the future has led some researchers to suggest that perhaps the main role of the episodic memory system is not to remember the past but to enable people to simulate possible future scenarios to help anticipate future needs and guide future behavior. After all, although the future hasn't happened yet, it will be happening pretty soon! And when the future becomes the present, we need to be able to act effectively. Information from the past, therefore, is a predictive tool that could be useful in deciding whether to approach or avoid a particular situation, both of which could have implications for effectively dealing with the environment and, perhaps, even for survival.[35] When conceptualized as a process that uses stored information from the past to construct scenarios that might happen in the future, memory becomes a tool for prediction.

Social Prediction

> Prediction in its many forms plays a fundamental and central role in social cognition.
> —Elliot Brown and Martin Brüne[36]

As we enter the world of social interaction, we encounter one of the most complex and significant areas of human behavior:

interacting with other people. Just imagine the advantage you would have if you could accurately predict what other people will do and why they are doing it. Consider, for example, the following situation:

> *Jim enters the conference room, nods at each of the people sitting around the table, glances at his watch, sits in one of the two empty seats, and whispers something to the person sitting next to him.*

What's going on here? Why does Jim nod to everyone and then glance at his watch? Is he late? Is someone else missing and he wonders where they are? Is he nervous because he has to make a presentation? You might be able to answer some of these questions based on what you know about Jim or about the situation leading up to this event, but it is likely that your answers will be a "best guess." Social scenarios, in which people observe other people or in which people interact with one another, are often filled with ambiguity and uncertainty.[37]

Basically, what we are trying to achieve in social situations is to understand why people are doing what they are doing and what is going on in their minds. The study of *social cognition*—the mechanisms by which people gather, store, and apply information about other people and social situations—is the gateway to understanding what was going on when Jim entered the conference room, as well as the multitude of social situations we encounter every day. Prediction is a central goal of social cognition, and researchers have approached the problem of social prediction in a number of different ways.[38] We begin by describing a central concept involved in explaining social prediction called *theory of mind*.

Theory of Mind (Mentalizing)

The problem of figuring out what is up with Jim as he enters the conference room comes under the heading of theory of mind, which David Premack and Greg Woodruff first described in 1978 by stating that "an individual has a theory of mind if he imputes mental states to himself or others."[39] They went on to explain that because we can't directly observe someone else's metal states (see chap. 1), we need to create a *theory* or make an *inference* about what their mental state might be.

Most of the early research on theory of mind focused on young children and asked (1) "Do young children have a theory of mind?" and (2) "When do children acquire a theory of mind?" In a typical experiment, children between the ages of about three and six years old are presented with the *Sally-Anne task*, which involves two protagonists, *Sally* and *Anne*. In one of the earliest Sally-Anne experiments, Simon Baron-Cohen and coworkers showed a child two dolls, one named Sally, the other named Anne.[40] The child sees the following play unfold: (1) doll Sally places a marble in a basket as doll Anne watches; (2) doll Sally leaves the room; (3) after doll Sally leaves, doll Anne moves the marble from the basket to a box; (4) when doll Sally returns, the experimenter asks the child, "Where will Sally look for the marble?" The typical finding is that the answer depends on the child's age. Children younger than three or four say that Sally will look in the box, where the marble had been moved, whereas children older than five or six say Sally will look in the basket, where the marble was before Sally left the room. These older children are said to have a correct theory of Sally's mind, because they realize that since Sally was not in the room when the marble was moved, she would not know that Anne had moved the marble to the box.

The Sally-Anne task, and others like it, are called *false-belief tasks*, because a person who answers incorrectly holds a false belief about what another person (or doll representing a person) knows. In research investigating theory of mind in adults, the term *mentalizing* is often substituted for, or used interchangeably with, theory of mind, although some researchers make subtle distinctions between the two terms.[41]

Mentalizing refers to the processes involved in dealing with social situations, and one of the tactics of research on mentalizing is to compare brain activity while participants read stories that involve mentalizing to activity that occurs in response to nonmentalizing stories. For example, consider two situations: A rock is rolling down a hill, or a person is running down a hill. When you see the rock rolling down a hill, you don't think, "It wants to get to the bottom." It is likely, however, that when you see a person running down the hill, you might make a prediction as to why the person is running. Because there is a bus stop at the bottom, one reasonable inference is that the person is running to catch a bus. And that prediction is strengthened when the bus appears, and the person starts running faster.

In an early paper in which participants read *mentalizing stories* (stories involving people) or *mechanical inference* stories (stories involving mechanical devices), Rebecca Saxe and Nancy Kanwisher had participants read the stories while having their brains scanned. What follows are examples of two of these stories.[42]

Mentalizing Story A boy is making a papier-mâché project for his art class. He spends hours ripping newspaper into even strips. Then he goes out to buy flour. His mother comes home and throws all the newspaper strips away.

Mechanical Inference Story A pot of water was left on low heat yesterday in case anybody wanted tea. The pot stayed on the heat all night. Nobody drank tea, but this morning, the water was gone.

The mentalizing story involves the mother having a false belief, since she does not realize that the newspaper strips are part of the boy's project. The mechanical inference story involves making the inference that the water boiled away overnight.

Saxe and Kanwisher focused on an area spanning the temporal and parietal lobes, called the *temporal parietal junction* (TPJ), because previous research had implicated this area in social cognition. Figure 5.6 shows the location of the TPJ plus the prefrontal cortex (PFC), which is also involved in social cognition.[43] The activity measured in the TPJ was clear-cut, with the response to the mentalizing activity being over three times greater than the response to the mechanical inference activity.

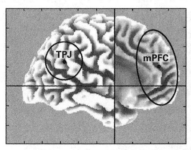 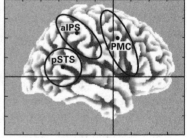

Figure 5.6
Left: Location of some structures in the "mentalizing network": TPJ = temporal parietal junction; PFC = prefrontal cortex. Right: Location of some structures in the "mirror network": STS = superior temporal sulcus; IPS = intraparietal sulcus; PMC = premotor cortex. There are also mirror neurons in other areas of the brain. (In contrast to other brain depictions in this book, the frontal lobe is on the right here.)

Two other conditions, in which people viewed photographs of bodies or objects, generated only small responses, indicating that the mentalizing responses were not just a response to the presence of a person but had to involve inferring what a person was thinking. Other research has confirmed the importance of the TPJ in theory of mind and social cognition in general and has identified a number of other areas including the prefrontal cortex and others that are also involved in mentalizing, all of which together are called the mentalizing network.[44] Researchers have found that when people have a large number of people in their social network, they are likely to have a large mentalizing network.[45]

Although some researchers identify the mentalizing network as the primary brain system involved in social prediction, others argue that an additional network, called the *mirror neuron network*, is also involved.

Inferring Intentions from Actions

While mentalizing research has focused on measuring people's reactions to different situations, like the Sax and Kanwisher experiment, another line of research has focused on how people draw conclusions about other people by observing their actions. We begin with the idea that observing actions can lead to conclusions about properties of the actor with a classic experiment by Fritz Heider and Marianne Simmel in which participants viewed short animated films that had three main characters: a small circle, a small triangle, and a large triangle (figure 5.7).[46] These characters moved in and around a small box, sometimes interacting with one another. As people watched these actions, they began describing what was happening not in terms of movements of geometrical objects, but in terms of an unfolding drama. The small triangle and circle were described as a couple, who just want to live in peace inside the box, but the mean big

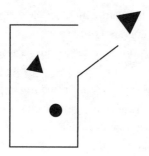

Figure 5.7
A still from the animated film used by Heider and Simmel. In the film, the three "characters"—Big Triangle, Small Triangle, and Circle—moved around and sometimes interacted with each other.

triangle is harassing them. Luckily, the small triangle is smart enough to close the door so the big triangle can't get in, at which point the big triangle throws a temper tantrum.

This is not Academy Award material, but it does illustrate something striking: people can attribute intentions and emotions to geometrical objects based on their movements. Experiments conducted over fifty years later showed that watching displays involving interacting geometrical objects like those in the Heider and Simmel experiment causes activation in areas of the brain associated with social cognition.[47]

Of course, our main concern is not the attribution of intentions and feelings to moving geometrical objects, but the attribution of intentions to humans. Imagine, for example that you see a person across the street waving their arm. How can you tell whether they are hailing a taxi or swatting a fly? The answer is that there are often subtle differences in movements that indicate the intention behind a movement. An experiment by Atesh Koul and coworkers had participants observe films of a hand reaching for a bottle, with the intention of either drinking from it or pouring from it.[48] Even though the film stopped just as the hand reached the bottle, the participants were still able to indicate whether the action that followed was going to be "pour" or "drink." They did this by noticing subtle differences in *visual kinematics* between the pour and drink conditions, where visual kinematics are characteristics of the hand's movement, such as velocity, trajectory, and nature of the grip.

While the kinematics of movement and grasping can suggest what a movement means, everyday life provides additional information in the form of the context in which the movement occurs. For example, in an experiment by Mario Iacoboni and coworkers, participants observed a film showing a hand reaching for a cup that is full of tea and is surrounded by food and other items that typically appear on a dining room table at the beginning of a meal, or a film in which the hand is reaching for an empty cup that is surrounded by the mess associated with the end of a meal. The answer to the question "why is the person reaching for the cup" is "to drink" in the first situation, and "to clean up" in the second. Context, in this case, provides cues as to the "why" of the action.[49]

The fact that movement cues and context cues can help people infer the "why" of an action isn't surprising. But the most important thing about these experiments is the physiological responses to the different films. To understand the motivation behind the physiological part of these experiments we need to go back to 1991 to a laboratory at the University of Parma in Italy, where researchers were studying how neurons in the monkey's motor area responded when the monkey carried out actions.[50] They located a neuron that responded when the monkey reached for and picked up a piece of food that was resting on the surface of a table. This result was interesting, but what happened next was extraordinary.

When an experimenter who was cleaning up after the experiment picked up a piece of food that was still on the table, the same neuron fired. Further research identified other neurons in the cortex that fired both when the monkey *carried out an action* and when the monkey *watched a person carry out the same action*. These neurons came to be called *mirror neurons*. Since these initial studies, a vast amount of research has studied mirror neurons in monkeys and has identified areas of the brain that have mirror properties in humans (figure 5.6b).

It has been suggested that one of the roles of mirror neurons is to help us understand other people's feelings. When a person's arm is touched, receptors in their skin send signals to neurons in the brain, which causes firing in neurons in the somatosensory cortex. But the somatosensory cortex is also activated when a person watches someone else being touched or watches someone else who is feeling pain.[51] This "firing to watching," which is associated with mirror neuron activity, has caused some researchers to suggest that neurons in these areas provide a mechanism for understanding other people's experience, and might, therefore, play a role in predicting how they might behave.

Returning to the visual kinematics experiment, when Koul and coworkers scanned their participant's brains as they watched the pour and drink actions, they found differences in the response to these two actions in areas that are part of the human mirror neuron system.[52] Similarly, in the reaching for the teacup experiment, Iacoboni and coworkers found that the mirror neuron system responded more strongly to "reaching to drink" than to "reaching to clean."[53] This result, according to Iacoboni, provides evidence that these neurons are responding to different intentions, and so are functioning to code the "why" of the actions. The neurons, according to this idea, are making a prediction of what is going to happen.[54]

How accepted are the two different neural approaches to social cognition: measuring mentalizing responses and recording activity in the mirror neuron network? There is wide agreement that the mentalizing system is important for determining intentions, but some controversy about mirror neurons. Many researchers have hailed mirror neurons as a way to understand other people's actions and feelings, but others question the claim that mirror neurons are involved in understanding the intention behind actions.[55]

One of the reasons for questioning whether mirror neurons signal intentions is that mirror neurons respond so rapidly after an action is observed that there may not be enough time to fully understand the action.[56] Concerns such as this led to the proposal of the *dual-process theory of mentalizing*, which states that the mirror neuron system may be involved in the rapid detection and recognition of actions, followed by the slower mentalizing system, which takes into account components of social behavior that are involved in determining a person's beliefs and intentions.[57] The continuing discussion regarding the role of mirror neurons in understanding social behavior aside, there is no question that large areas of the brain are involved in helping people make predictions about the beliefs and goals of others.

The Multiple Faces of Prediction

That we can describe all the widely varied functions summarized in table 5.1 as involving predictive mechanisms argues that prediction is a central mechanism of the mind. But to fully appreciate the nature of prediction, it is necessary to realize that although

Table 5.1
The Many Faces of Prediction

	Phenomena	Mechanisms
Perception	Visual perception	Unconscious inference (Helmholtz). Likelihood principle, determined by past experience.
Motor response	Eye movements (guidance)	Guided by past experience that internalizes statistics of the environment.
	Eye movements (cause image smear)	Corollary discharge predicts movement and prevents perceiving image smear.
	Reach and grasp	Governed by *how* pathway. Aided by information from *what pathway*.
	Tickling yourself	Difficult because of presence of corollary discharge.
Language	Speech segmentation	Based on experience with the statistics of language and sentence context.
	Predicting upcoming words	Cloze probability test: some words more likely in a particular sentence. Physiology. Brain anticipates what is most probable.
	Garden path sentences	Incorrect parsing is influenced by word meaning, context, past experiences.
Music	Predicting timing	"Taking up the beat." Syncopation illustrates less predictable situation.
	Predicting notes	Measured by musical cloze probability and monitoring response to unlikely chords.
Memory	Construction of memory	Bartlett experiment. Memory changing over time. Memory as construction.
	Predictive editing	Memory trace is weakened when a prediction is not fulfilled.
	Simulating the future	Role of episodic memory in anticipating the future.
Social	Theory of mind (mentalizing)	Understanding motivations behind another person's behavior and what they are thinking.
	Mirroring actions	Mirror neurons as involved in understanding others' feelings and motivations.

on one level prediction is an overriding principle, prediction is not a monolithic mechanism that operates in the same way across all its functions. Looking across different functions reveals that prediction occurs at different timescales, involves different mechanisms, and is accompanied by different degrees of awareness.

The timescales of prediction range from milliseconds, when considering the speed of perception, to seconds, as prediction guides a person's reach for an object or their understanding of a sentence, to minutes or longer, as a person determines how to interpret other people's social behaviors or as a conversation unfolds.

With regard to social behaviors, Jorie Koster-Hale and Rebecca Saxe point out that the social environment can be predicted at a range of timescales, ranging from milliseconds (where will the woman look when the car honks?) to minutes (will the woman look in the usual places for her keys when she is ready to leave?) to months (what would it be like to be in a relationship with the woman?).[58]

Given the different timescales of prediction, we should not be surprised that these different instances of prediction may involve different mechanisms. For example, sensorimotor processes occur as the corollary discharge associated with motor signals modulates what we perceive as we move our eyes or determine the strength of our grip. However, more cognitively based processes are involved in understanding sentences and appreciating music, simulating the future, and decoding social situations.

Finally, a continuum of awareness is associated with these different mechanisms. The operation of the corollary discharge operates automatically and beneath awareness, as in many cases does predicting upcoming words in a sentence or anticipating sentence parsing. But other processes, such as predicting what might happen in the future or determining a theory of another person's mind, are more likely to operate on a conscious level.

Thus, looking at the many functions of prediction, we see that it is an overarching principle of the mind's functioning that is served by numerous individual mechanisms, all working toward a common goal: ensuring our survival as we negotiate our way through the environment.

Beyond Prediction

Chapter 4 introduced the hypothesis that people strive to reduce prediction error. An implication inherent in this idea is that prediction is good, and error is a bad thing that must be avoided or eliminated. Applied to music, this idea translates into how the beauty of music can be enhanced by our ability to predict what is coming next in a composition, as when there are repeating themes.

But we have also seen, in the discussion of the benefits of syncopation in music, that prediction can serve another purpose. Prediction creates a template to be violated, and sometimes violating this template can be pleasing, as when syncopated music induces us to dance, or when other unpredictable events create novelty and surprise, which some have argued are basic human needs.[59] So to go with the saying "the brain is a prediction machine," we can add another one: "novelty is the spice of life."

It is no coincidence that the goal of many writers and artists is to create art that defies our predictions. Thus writers often challenge our predictive mechanism with stories that lead us down the garden path and may include problem-solving ("Who did it?"). Architects sometimes create buildings that do not reveal their secrets when seen from a single viewpoint, a notable example being the Sydney Opera House (fig. 5.8).

Figure 5.8
The Sydney Opera House from different viewpoints. Beginning with the view on the bottom right, moving clockwise around the pictures reveals views that would be seen from a boat sailing clockwise around the Opera House. More than one viewpoint is necessary to understand this structure. (Photo credit: Bruce Goldstein.)

Composers often purposely insert unexpected notes or chords in their compositions to create surprise or interest.[60] Wolfgang Amadeus Mozart often used this device. For example, in an opening phrase of his thirty-first symphony there is a rising scale that begins on a D and ends on a D. This ending is highly expected, so listener's familiar with Western music would predict the D if the scale were to stop just before it was to happen. But something different happens later in the movement, in which the first notes are A's, which again are followed by a rapidly rising scale. But at the end of the scale, when listeners expect an A, they instead hear a B-flat, which doesn't sound like it fits the note predicted by the return to the tonic. This lack of fit is not, however, a mistake. It is Mozart's way of saying to the listener "Listen up. Something interesting is happening!"

So we end our description of prediction with a caveat. Yes, prediction is a good thing, which keeps us on the right track and is highly adaptive. However, sometimes it's nice to be surprised, to experience novelty, to have our interest piqued by the unexpected. So the brain, in all its wisdom, makes predictions that both help guide us through life and can be violated to give us pleasure and satisfy our need for novelty.

6 Dynamic Highways of the Mind

How does the brain create the mind and therefore experience? The discussion of the creation of experience in chapter 2 described the "hard problem of consciousness," which many researchers have declared may be unsolvable. Instead, they have focused on the "easy problem of consciousness," which involves determining neural correlates of consciousness—looking for connections between neural responding and experience.

An important event in the search for the neural correlates of consciousness occurred in the nineteenth century when Broca's and Wernicke's studies on people with brain damage identified brain areas that are specialized for language. The twentieth century brought with it new technologies that made it possible to record the activity of neurons using single-neuron recording, and to determine patterns of brain activation using brain imaging techniques such as fMRI. Throughout the twentieth century, mountains of evidence accumulated for localization of function—there are brain areas that respond best to specific types of stimuli—and distributed processing—a particular stimulus can activate many areas distributed throughout the brain.

Looking at brain functioning in terms of localization of function and distributed processing focuses on *where* stimuli are processed in the brain. But understanding the brain is about more than *where*. It is also about *how*. The two chapters on prediction were very much about *how*, as they discussed how neural signals traveling up from the receptors interact with neural signals traveling down from the brain to create information that enables us to make predictions.

This final chapter continues the story of *how* by considering the dynamics of activity that occurs as neural signals travel to various places in the brain. We do this by conceptualizing neural signals as traveling throughout the brain in a vast neural highway made up of large networks of neurons. The operation of this highway system comes under the heading of "breaking news," because it is one of the current hot topics in the study of the relationship between mind and brain. But even breaking news has ties to the past, so we begin in 1942 with Sir Charles Sherrington's eloquent description of the brain as an enchanted loom.[1]

> The brain is waking and with it the mind is returning. It is as if the Milky Way entered upon some cosmic dance. Swiftly the head mass becomes an enchanted loom where millions of flashing shuttles weave a dissolving pattern, always a meaningful pattern though never an abiding one; a shifting harmony of subpatterns.

The passage comes from Charles Scott Sherrington's book *Man on His Nature*, published ten years after he received the 1932 Nobel Prize in physiology and medicine (with Edgar Adrian; see chap. 2) for his pioneering research on how neurons communicate. The beginning of the passage refers to the brain waking up from sleep and is based on the incorrect assumption that the mind is not active during sleep. But we can forgive this inaccuracy, because what follows is a poetic description of the mind at work, with the most famous phrase describing the brain as "an enchanted loom." The idea of an enchanted loom has resonated with modern writers, who have incorporated it into the titles of their books, such as

Robert Jastrow's *The Enchanted Loom: Mind in the Universe* (1981) and Pietro Corsi's *The Enchanted Loom: Chapters in the History of Neuroscience* (1998).[2] It is clear from the imagery of phrases such as "cosmic dance," "floating shuttle," and "shifting harmony of subpatterns" that Sherrington envisioned the brain's enchanted loom as being dynamic and constantly changing.

Sherrington's ideas, and other people's research in the decades that followed, confirmed both the dynamic nature of the brain and that the activity is distributed throughout the brain. This picture of the brain as dynamic, with activity distributed throughout, is one we have encountered as we have looked at changes unfolding over time ranging from movement of the eyes, to reaching for a bottle, to understanding language, listening to music, and interacting with other people. We begin by describing the structure of the brain's communication system.

The Brain's Communication System

> Network maps are fundamental for understanding the brain's structural and dynamic organization.
> —Olaf Sporns[3]

What is a network map? The quote above points to two types of organization, structural and dynamic. We begin with the structural map called the connectome.[4]

The Connectome

The connectome is the "structural description of the network of elements and connections forming the human brain,"[5] or more simply the "detailed 'wiring diagram' of the neurons and synapses in the brain."[6] Mapping the connectome has been a central goal of researchers, as evidenced by the Human Connectome Project, which was funded in 2010 with a grant of $40 million from the National Institutes of Health.[7]

Mapping the connectome, however, is no simple task, as the brain consists of tens of millions of neurons, each making connections with thousands of other neurons. One method for determining this "road map" of the brain, called *track-weighted imaging*, is based on detection of how water diffuses along the length of nerve fibers (fig. 6.1).[8] The connectome's wiring diagram

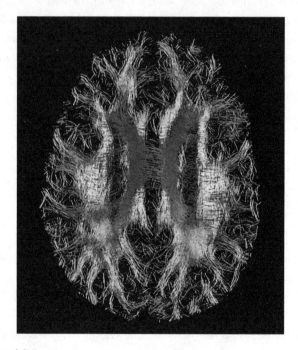

Figure 6.1
The connectome. Nerve tracts in the human brain determined by track-weighted imaging. The color version of this figure more clearly differentiates the various nerve networks.

is known as the map of *structural connectivity*, but to understand how electrical signals travel through this network, we need to introduce the idea of functional connectivity.

Functional Connectivity and Functional Networks
Picture the road network of a large city. On one set of roads, cars are streaming toward the shopping area just outside the city, while on other roads cars are traveling toward the city's business and financial district. One group of people is using roads to reach places to shop; another group is using roads to get to work or conduct business. Just as different parts of the city's road network are involved in achieving different goals, so different parts of the brain's neural network are involved in carrying out different cognitive or motor goals.

The quote that opened our discussion of localization of function in chapter 2 used the term *module*, which is an area specialized for a specific function. Some researchers use this term, and others use the term *functional networks*, because locations within these areas are functionally connected to each other. The important thing about these areas, whether they are called modules or functional networks, is that locations within an area are functionally connected to one another. We can understand what being functionally connected means by considering how functional connectivity is measured.

One way of measuring functional connectivity is the *resting-state method*, in which fMRI is used to measure resting-state activity—activity when the brain is not involved in a task—in a reference location, like the circled location (a) in figure 6.2.[9] When activity measured at other locations is compared to the activity at the reference location, some activity is highly correlated, as indicated by the numbers over the records, which means

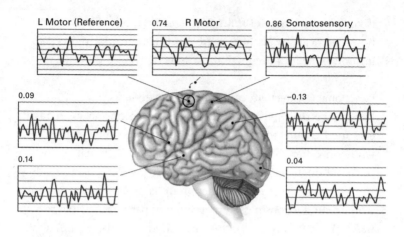

Figure 6.2
Determining functional connectivity with resting-state fMRI. The resting-state activity measured by fMRI is shown for the reference location in the left motor cortex, and for six other locations in the brain. The right motor location is on the other side of the brain. The numbers above each record indicate the correlation with the reference location. Correlations for the right motor cortex (0.74) and somatosensory cortex (0.86) are high, indicating good connectivity. These three locations are part of the somato-motor network. The other locations are not highly correlated and thus are outside of the network.

that the ups and downs of the waveforms, if superimposed, overlap. High correlations occur for R Motor and Somatosensory, but not for the other records. Thus, L Motor, R Motor, and Somatosensory are functionally connected with the strength of the connections indicated by the size of the correlation.

Figure 6.3 shows brain areas for six of the brain's functional networks, with functions described in table 6.1.[10] Identifying networks that serve different functions is an important first step in understanding the brain's communication system. But to really understand what is happening during cognition, we need to go

Dynamic Highways of the Mind

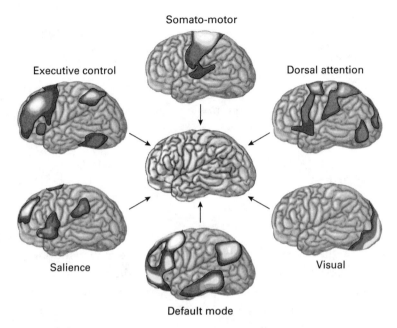

Figure 6.3
Six major brain networks determined by the resting-state method. Note that all these networks increase activity in the highlighted areas during a task and decrease activity when at rest, *except* the default mode network, which decreases activity during a task and increases activity when there is no task. See table 6.1 for brief descriptions of these networks.

beyond just identifying areas that serve different functions. We need to consider the how information flows in these networks.

Note that there are other networks as well, including networks involved in hearing, memory, and language.

Information Flow within and between Networks

To understand what we mean by information flow, let's return to our analogy between the structural map of the brain and a big-city street system. Imagine climbing into a helicopter and flying

Table 6.1
Six Common Functional Networks Determined by Resting-State fMRI

Network	Function
Visual	Vision; visual perception
Somato-motor	Movement and touch
Dorsal attention	Attention to visual stimuli and spatial locations
Executive control	Higher-level cognitive tasks and directing attention during tasks
Salience	Attending to behaviorally relevant events
Default mode	Mind wandering and cognitive activity related to personal life story, social functions, future simulation, and creativity

above the city so that you can observe the patterns of traffic flow at various times of day. As you hover above the city, you notice how this flow changes when the street system is serving different functions. During morning rush hour, when its function is to get people to work, there is heavy flow from the suburbs toward the city on the major highways. Evening rush hour reverses the flow on the major highways as people head for home, and the flow may also increase on suburban streets a little later. During the day, traffic flow may be higher around shopping areas; and before and after special events, like a weekend football game, flow will be high on roads leading to and from the stadium.[11]

The point of this example is that just as traffic flow in the city changes depending on the functions being served by the road system, so the flow of activity both within a functional network and between different functional networks changes, depending on which functions are being served. For example, consider what happens when a person looks at a cup of coffee on a table. Looking at the cup causes activity in the visual functional network as the person perceives the various qualities of the cup.

Meanwhile the dorsal attention network may also be activated as the person focuses attention on the cup, and then the somatomotor network is activated as the person reaches to pick up the cup, grasps it, and lifts it to drink. During this process, the executive control network is actively monitoring and coordinating the capacities of attending, reaching, and lifting. So even a simple everyday experience like looking at and picking up a cup of coffee involves activity within individual networks, plus rapid switching and sharing of information between a number of different functional networks.[12] The remainder of the chapter will focus on research that has determined the relationship between functional networks and cognition.

Representing Networks
Once functional networks have been determined, how do we represent the network in a way that provides more detail than the shaded areas in figure 6.3? There are many different ways of representing networks.[13] *Graph theory* represents networks in a way that is easy to visualize by mapping networks as *nodes* connected by *edges* (fig. 6.4a). Nodes are functional processing units, such as neurons, groups of neurons, or brain areas. Edges represent interactions between the nodes.

Two ways to graphically represent networks are shown in figures 6.4b and 6.4c.[14] These depictions show connections between different modules (functional networks). The network in figure 6.4b is plotted in *anatomical space* on the brain, with each module represented by a node placed at its actual location on the brain. In this example, the nodes are large areas, like the visual area and the prefrontal area. The strength of the connection between areas is indicated by the thickness of the edges. The thick edge in figure 6.4b indicates a strong connection between the visual area, indicated by the dark circle, and the prefrontal area, indicated by

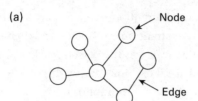
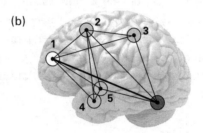
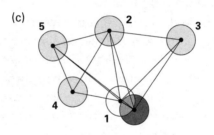
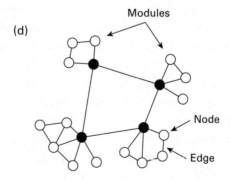

the light circle. The network in figure 6.4c is plotted in *functional space*, with the location of the nodes determined by how related they are to each other. In this depiction, closeness of the visual and prefrontal areas indicates that they are strongly connected.

We can use graph theory not only to indicate connections between modules but also to indicate connections within modules. The network in figure 6.4d illustrates connections between four different modules, as well as some of the connections within each module.

With the basics of network graphs in hand, we are now ready to look at the dynamics of cognition, as pictured by how networks change under different conditions.

The Dynamics of Cognition

An amazing thing about functional connections is that they change depending on conditions. In the remainder of the chapter, we will look at a number of examples of how these changes occur. The first example involves the medial temporal lobe (MTL), an area involved in memory storage (see fig. 2.1).

Figure 6.4
(a) A simple graphical network showing nodes and the edges that connect the nodes. (b) Functional connectivity between areas, shown in anatomical space, with their locations corresponding to their locations on the brain. The strength of connection is indicated by the thickness of the edges. (c) Functional connectivity shown in functional space. The strength of connection is indicated by the distance between nodes, with shorter distances indicating stronger connections. (d) A functional network that contains four modules, each of which serves a specific function.

Shifting Connections
Just as traffic flow on the highway changes from morning to evening, so neural traffic to the MTL is different in the morning and the evening. In the morning, most MTL activity is localized within that structure, but in the evening, connections carry signals out to other areas of the brain.[15] Why does this happen? One idea is that connectivity between areas increases as people accumulate experiences during the day, and then decreases at night as memories are strengthened by a process called memory consolidation. By morning, the MTL is sending signals mainly within itself, and the process begins again. Thus connectivity is not static but varies over the course of the day.

Another indication of the changeability of functional connections is demonstrated by an experiment in which functional connectivity of one person's brain was determined from 100 brain scans carried out over 500 days.[16] Figure 6.5 compares connectivity on two of those days: Tuesday, after fasting and no caffeine; and Thursday, after eating and having caffeine.

Two changes stand out in the pictures in figure 6.5. In the fasting condition, the visual and somato-motor modules are closely linked, whereas they are not functionally connected in the fed condition. Also, notice what is going on inside the somato-motor module. There is lots of clustering in the fasted condition, but much less communication between nodes in the fed condition.

Although it is unclear what the changes that occur between fed and fasted mean, the important point for our purpose is that changes do occur, and they are not subtle changes. Changes in physiology associated with eating or drinking coffee cause large shifts in the traffic pattern in the brain as connectivity changes within and between modules.

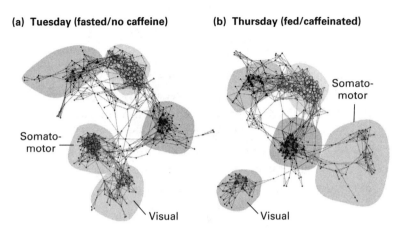

Figure 6.5
Effects of eating/caffeine on functional connectivity structure. (a) Network measured on Tuesday, after fasting and no caffeine; (b) Network measured on Thursday, after eating and caffeine. The somato-motor and visual modules are mentioned in the text.

Interacting Networks

We have seen how the traffic patterns within the brain can change throughout the day as memories are processed and across many days in response to physiological changes. But when we look at ongoing behavior, we are confronted with changes that occur rapidly, on timescales of seconds or minutes. For example, consider our example of reaching for and picking up a coffee cup. Or consider what happens when you are driving and the traffic light turns red. You see the change (visual area) and slam on the brake (motor area). Are actions such as these, which involve associations between brain areas, accompanied by changes in connectivity? We answer this question by focusing on two types of ongoing behavior: creative thinking and social interactions.

To tell this story, we return to the networks in figure 6.3 and consider something interesting about one of them.

The Default Mode Network and Mind Wandering

Five of the networks in figure 6.3 behave as we would expect: they increase their activity when engaged in their task (see fig. 6.3, table 6.1). But one of these networks, named the *default mode network* by Marcus Raichle and coworkers in a 2001 paper titled "A Default Mode of Brain Function," behaves differently.[17] Activity in the default mode network (default network for short) decreases when a person is engaged in a task, but it increases when the mind is at rest.

What does activity "at rest" mean? One attempt at answering this question focused on a phenomenon that often accompanies activity in the default network: the mind wanders. Initial research on mind wandering cast it in a negative light, as evidenced by the title of Matthew Killingsworth and Daniel Gilbert's paper "A Wandering Mind Is an Unhappy Mind."[18] The basis for this conclusion was their study in which over five thousand people used an iPhone app that contacted them at random times and asked them about the thoughts, feelings, and actions they were experiencing. People's minds were wandering about 47 percent of the time, and mind wandering was, on average, associated with negative mood.

Mind wandering also had other negative effects, linked to the fact that mind wandering is also called *task-unrelated thought*. Clearly task-unrelated thoughts are not a good thing if they accompany a task that requires being focused and therefore not distracted. In line with this idea, mind wandering has been associated with decreases in performance of tasks such as driving, reading, and paying attention to ongoing events.[19] These negative effects of mind wandering led to negative attributions about the default network. After all, it is associated with mind wandering, and because it is active when the brain is "at rest," it came to be called the "task-negative network."

But it did not take long for researchers to discover the positive side of mind wandering and the default network. Benjamin Baird and coworkers conducted an experiment that connected mind wandering and creativity, based on the observation that when a person is working on a problem but can't solve it, the solution sometimes "appears" after the person has put the problem aside.[20] This phenomenon has been noted by scientific thinkers, including Albert Einstein, Henri Poincaré, and Isaac Newton, who have described having moments of inspiration when they had stopped thinking about a problem they had been trying to solve. The phenomenon of getting ideas after taking a "time-out" from working on a problem is called *incubation*.

Baird's experiment began with a baseline task, called the *alternate uses task* (AUT) (also called the *unusual uses task*), in which participants had two minutes to think of unusual uses for common objects. For example, how many unusual uses can you think of for bricks? (A few examples: use as a weapon, a paperweight, a stepping-stone, an anchor.)

The baseline AUT was followed by a twelve-minute incubation period, during which participants carried out a difficult task, which resulted in a low rate of mind wandering, or an easy task, which resulted in a higher rate of mind wandering. When participants then repeated the AUT for the same objects they had considered before, the results were clear-cut: after the easy task, which was accompanied by a high rate of mind wandering, performance on the repeat AUT increased by 40 percent compared with the baseline. After the hard task, performance was unchanged. Mind wandering, Baird concluded, facilitates creative incubation.

In addition to the connection between mind wandering and creativity, other positive effects of mind wandering were discovered. For example, people often mind wander about social situations, and research has found that social mind wandering often leads to positive feelings and increased happiness.[21] In addition, evidence suggests that mind wandering can help us plan for the future, a process called *autobiographical planning*.[22]

Meanwhile, back at the default network, a similar story of initial negativity followed by redemption was unfolding, because many of the positive effects of mind wandering were accompanied by default activity. In addition, mind wandering is associated not just with default network activity but with activity in the executive control network, which is involved in directing attention during a task.[23] This finding raises an interesting question: how can two networks that are "opposite"—one that is active when a person isn't focused on a task, and another that is active when a person is carrying out a task—coexist at the same time? A hint at the answer to this question has come from research on creative thinking.

Creative Thinking

> Because the brain is a highly complex system composed of functionally interconnected neural networks, the interaction between individual regions is crucial to understanding how cognitive processes like divergent thinking unfold.
>
> —Roger Beaty et al.[24]

Creative thinking is a wide-ranging topic, which involves looking at problems and situations from a new perspective and includes activities like writing, creating art and music, inventing new devices, and solving puzzles and math problems.[25] We will focus on *divergent thinking*, which is a central component of general creative ability and has stimulated research that teaches us something important about how the mind operates.

Divergent thinking is the process of generating creative ideas by exploring many possible solutions. A common method of measuring divergent thinking is the alternate uses task by which Baird demonstrated the link between mind wandering and creativity. What makes divergent thinking particularly relevant to our quest to understand the link between networks and the mind is that two "opposing" networks—the default network and the executive control network—are both activated during creative thinking. An experiment by Roger Beaty and coworkers sheds light on this seemingly unlikely pairing.[26]

Beaty first determined how the functional connectivity between these two networks varies in people with high and low creative ability.[27] Based on performance on the alternate uses task, Beaty and coworkers divided participants into high-creativity and low-creativity groups. When they measured functional connectivity using the method illustrated in figure 6.2, the researchers found that the high-creativity group had a stronger connection between the default network and the executive control network.

What this connection suggests is that creative thinking may involve cooperation between these two "opposing" networks. Perhaps, Beaty suggests, the default network is important for generating ideas and the executive control network for evaluating ideas—weeding out unoriginal ideas and highlighting original ideas that are more likely to contribute to creativity.[28]

This idea is supported by the results of an experiment in which activation was measured as participants designed book covers. As in the alternate uses task, the default network was more active as participants were generating ideas for the covers, and the executive control network was more active when participants were evaluating the ideas.[29]

The idea that two networks can serve different functions for the same task helps explain the role of the default network in creativity. But perhaps even more important for our purposes, it provides an example of how networks interact. In the case of creativity, active cross talk occurs between two networks with different functions, which are usually active at different times but can get things done if they cooperate with each other.[30]

Beaty also considered what happens as creative thinking unfolds over time. He used a technique called *temporal connectivity analysis* to track second-by-second changes in connectivity just after a new object was presented in the alternate uses task.[31] Figure 6.6 plots connectivity between an area within the default network (black dot) and areas outside the default network (white dots). In the first two seconds of thought, the default network connects with three other areas, followed by connections to other areas during seconds four and six, and then fewer areas at second eight. Two of the key areas that connect with the default network are the salience network and the executive control network, both of which are involved in controlling attention. Thus the dynamic nature of thinking is accompanied by rapid changes in network connectivity, with at least some of these changes presumably related to the cooperation between the default network and the executive control network.

The idea of coordination between networks is especially relevant to another function of cognition that is dynamic and active and potentially involves lots of switching: thinking about interacting socially with other people.

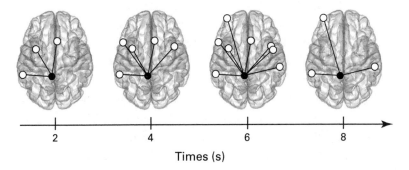

Figure 6.6
Dynamic changes in functional connectivity, beginning two seconds after person is presented with an object word (like "brick") in the alternate uses task. The black dot is a location in the default network. The other dots are located in various other networks outside the default network. It is clear from these plots that rapid changes in connectivity take place over an eight-second period.

Social Cognition

The discussion of social cognition in chapter 5 noted that because we can't directly observe someone else's mental state, we have to infer what that mental state might be. We also saw that the mentalizing network, which includes the temporal parietal junction and the prefrontal cortex, is involved in making these inferences. It is here that we pick up the story of social cognition, by considering an experiment by Ralf Schmälzle and coworkers that studied how the structure of the mentalizing network is affected by a social situation.[32]

The social situation that Schmälzle studied occurs when a person is excluded from a group. One way social exclusion has been created in the laboratory is by having a participant play a video game called *Cyberball*. When Schmälzle's participant arrived for the experiment, he or she met two other people, who, the participant was told, were also going to participate in the experiment. All three were assigned to separate booths and sat in front of a computer screen, where their task was to control a cartoon person on the computer screen so that it "played catch" with the cartoon people controlled by the other two participants (fig. 6.7).[33]

As the three cartoon people began throwing a ball back and forth, everything was fine for a while, with everyone getting the ball equally. But suddenly the other two people began throwing the ball back and forth, excluding the participant. Unknown to the participant, the other two cartoon people were controlled not by the two other "participants" (who were actually working with the experimenter) but by the computer. Participants in *Cyberball* experiments such as this one typically become upset and feel rejected when they are excluded.[34]

When Schmälzle had participants play *Cyberball* while he measured their functional connectivity in a scanner, he found

Figure 6.7
What the participant in Schmälzle's experiment saw in the monitor. The hand on the bottom represents the participant. Left: At the beginning, the ball is passed among all three characters. Right: Later, the participant is left out.

that social exclusion increased connectivity within the mentalizing network. One interpretation of this increase is that because the mentalizing network is involved in figuring out and reacting to social situations, the increased connectivity represents the participant's attempt to cope with being excluded. But the really interesting result is that the amount the connectivity increased depended on the structure of the person's social network—the network of the person's friends and acquaintances.

We can depict social networks in a graph similar to the ones we have been using for the brain, but with people at the nodes instead of neurons or structures, and with the distance between two people representing the closeness of their connection. One characteristic of social networks is *denseness*. A person has a dense social network when his or her friends are also friends with one another and so are positioned near one another in the network. A person has a sparse network when his or her friends don't know one another and so are spread out in the network.

When Schmälzle determined people's social networks by analyzing their "friends" on their Facebook pages, he found that people with sparser social networks were more likely to react with greater connectivity increases in their mentalizing network when they were excluded. Why might this occur? One possibility is that the structure of a person's social network may be saying something about how vulnerable a person is to being affected by social exclusion. Perhaps not having a close-knit group of friends makes it more difficult for people to deal with being excluded. This study is just one of many linking social cognition and brain connectivity.[35]

Another connection between social cognition and networks is that people who have larger social networks tend to have a larger prefrontal cortex, which is part of the mentalizing network.[36] Finally, when two closely related people interact, their brain waves are often synchronized, whereas the brain waves of unrelated people interacting are often quite different.[37] Although this synchronization research has so far focused not on *connectivity* but on *activity*, it is tempting to speculate that perhaps people whose brain waves are synchronized while they are interacting might also have similar brain networks.

Networks across the Life Span

> Performance in...cognitive functioning has been found to decline with age....There is evidence that these deteriorations are partly related to changes in conversation between different brain areas.
> —Linda Geerligs et al.[38]

Network connectivity changes within seconds or minutes during creative thinking[39] and dealing with social situations[40] or within hours when connected with memory storage[41] and reaction to physiological conditions.[42] Now we look at an even more extended timescale: the changes that take place over a person's lifetime.

Structural changes play a big part in the story of connectivity and aging. The literature documenting structural changes in the brain during aging is too voluminous to even summarize here, but two of the most important structural changes, which have been linked to declines in cognitive performance on tasks involving memory, processing speed, and attention, are (1) a decrease in the volume of the hippocampus, a structure crucial for memory,[43] and (2) a decrease in the number of pathways and connections.[44] Changes in structural connectivity are also associated with changes in functional connectivity, which is what we will focus on.[45]

Figure 6.8 shows functional connectivity graphs determined by Micaela Chan and coworkers[46] for young males (age 20 to 34) and old males (age 65 to 89). One major change that occurs as a function of aging is a spreading out of modules for the older participants. This means that there is less correlation *within* a module. Another change is a decrease in modularity. That is, the separation between different modules decreases, in some cases resulting in an overlap between modules that were separated in young people. This overlap has been called *dedifferentiation*, which refers to the loss of separation or differentiation between modules.[47]

These effects are illustrated in figure 6.8 by what is happening within the ellipses, which enclose two different modules, one with dark circles and the other light. For the old participants, the dark circles and light circles are farther apart, indicating poor communication within the modules.[48] There is also more overlap between the dark and light circles, indicating a decrease in modularity.

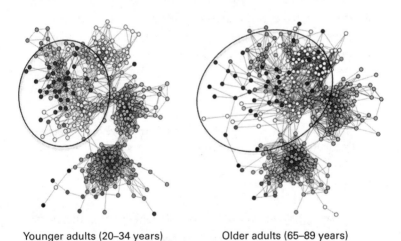

Younger adults (20–34 years)   Older adults (65–89 years)

Figure 6.8
Functional connectivity networks for (a) younger adults (20–34 years) and (b) older adults (65–89 years). Modules are identified by different shadings of the small circles, which show up better in the color version. The ellipses indicate how the areas occupied by two different networks are expanded in the older adults.

Chan also gave memory tests to each participant and found that less separation between modules was associated with poorer memory, no matter what the age of the participants. He found young participants with poor memory performance whose modules were close together, and older participants with good memory performance whose modules were separated. Something about modules being separated contributes to better memory.

So connectivity changes with aging, with an important effect being that modules tend to merge as people age, and this effect is associated with poorer performance on memory tests. This research doesn't answer why less separated modules should have this effect, but other research has begun looking at how aging affects the way different networks interact while people are carrying out tasks.[49] In addition, the National Institutes of Health connectome project has been expanded by adding the "Lifetime Human Connectome Project in Aging," which will create brain maps for people ranging from thirty-six to one hundred years or older![50]

What Does This All Mean?

> The specific brain characteristics that define an individual are encoded by the unique pattern of connections between the billions of neurons in the brain.
>
> —Fang-Cheng Yeh et al.[51]

This chapter has continued the story of brain function that began in chapter 2, which introduced localization of function (functions are located in specific areas of the brain) and distributed representation (cognitive functions activate a number of areas distributed across the brain). We have seen that to explain the connection between mind and brain, we need to understand how the many brain areas that create distributed representations are connected and how neural traffic travels along these connections. The story that has emerged has, in fact, shown that a person's characteristics are encoded by the patterns of connections between neurons. This is illustrated in the six studies listed in table 6.2, each of which illustrates a specific property of networks.

These studies provide a snapshot of many of the issues that connectivity researchers are thinking about today, and looking at all of them together, we can appreciate a new vision of brain functioning that extends beyond simply identifying areas that are activated by the brain, to understanding the constantly changing patterns of traffic that travels along the neural pathways that connect these areas.

The story of connectivity told in this chapter argues that structural and functional connectivity are important determinants of both who we are and how we behave and feel at any moment in time. But as exciting as this idea is, the story told here represents just the initial chapter in what promises to become a long novel about how the mind is created by the brain. One challenge for future researchers is to understand how to get from *encoding by patterns of connections*, referred to in the epigraph at the beginning this section, to *mechanisms* that translate these patterns of connections into specific cognitive functions. Thus we have seen that old age brings with it both (1) dedifferentiation (modules that are close together and sometimes even overlap, as in fig. 6.8) and (2) decreases in cognitive performance, but we do not yet understand the specific mechanisms that transform changed patterns of connectivity into decreases in cognitive performance.

These gaps in our knowledge aside, the story of research on the connection between mind and brain is an impressive one. Consider, for example, the statement by the Buddhist monk

Table 6.2

Changes in Functional Connectivity Described in this Chapter

Condition	Result	Basic Principle
Morning to evening (Shannon[1])	Medial temporal lobe connections increase during the day, decrease at night.	Functional connectivity changes throughout the day.
Fasting / no caffeine vs. Eating/caffeine (Poldrack[2])	Differences occur in distance between modules.	Functional connectivity is influenced by physiological conditions.
Creative thinking: divergent uses task (Beaty[3])	High-creativity people have stronger connection between default mode and executive control network.	Networks cooperate during thinking. Individual differences exist.
Creative thinking: divergent uses task, dynamic connectivity (Beaty[4])	Changes in connectivity occur during the first seconds of creative activity.	Network connections can change rapidly.
Social cognition: social exclusion (Schmälzle[5])	Increased activity in mentalizing network is associated with social exclusion. Effect is larger for people with sparse social networks.	Activity within a network can be affected by social stress. Individual differences exist.
Aging (Chan[6])	Less separation *between* modules is associated with aging (dedifferentiation). Fewer connections *within* modules with age.	Functional connectivity changes on a lifetime scale.

[1] B. J. Shannon, A. Desenbach, Y. Su, Y. et al., "Morning-Evening Variation in Human Brain Metabolism and Memory Circuits," *Journal of Neurophysiology*, 109 (2013): 1444–1456.

[2] R. A. Poldrack, T. O. Laumann, O. Koyejo, et al., "Long-Term Neural and Physiological Phenotyping of a Single Brain," *Nature Communications* 6.8885 (2015): 1–15.

[3] R. E. Beaty, M. Benedek, R. W. Wilkins, et al., "Creativity and the Default Network: A Functional Connectivity Analysis of the Creative Brain at Rest," *Neuropsychologia*, 64 (2014): 92–98.

[4] R. E. Beaty, M. Benedek, S. B. Kaufman, and P. J. Silvia, "Default and Executive Network Coupling Supports Creative Idea Production." *Scientific Reports*, 5, no. 10964 (2015): 1–14.

[5] R. Schmalzle, M. B. O'Donnell, J. O. Garia, et al., "Brain Connectivity Dynamics during Social Interaction Reflect Social Network Structure," *Proceedings in the National Academy of Sciences* 114, no. 20 (2017): 5153–5158.

[6] M. Y. Chan, D. C. Park, N. K. Savilia, S. E. Petersen, and G. S. Wig, "Decreased Segregation of Brain Systems across the Healthy Adult Life Span," *Proceedings of the National Academy of Sciences* 111, no. 46 (2014): E4977–E5006.

Geshe Kelsang Gyatso from chapter 1, "Our brain is not our mind. The brain is simply a part of our body that, for example, can be photographed, whereas our mind cannot."[52] While it is correct that we cannot photograph the mind, we have provided ample evidence that rejecting the centrality of the mind-brain connection is not correct. Yes, the mind and brain are not the same thing, but the brain creates the mind, and cognitive scientists have managed to study the mind, even though it can't be photographed. That they have been able to discover so much at the physical level—neurons, modules, networks—and relate those discoveries to the level of behavior and feelings is one of the most impressive achievements of modern science.

In closing, it is important to acknowledge that cognitive scientists have created many stories about the mind, in addition to the one I have told in this book. While I have focused on the basic principles of hidden processes, prediction, and connectivity, other books, some of which are listed in "Further Readings," provide greater detail regarding how each of the functions of the mind listed in chapter 1—perception, attention, memory, emotions, language, deciding, thinking, reasoning, and taking physical action—is created by the mind and the brain.

But greater detail aside, the stories told in those books have in common the idea that behind every cognitive function, the brain is working silently behind the scenes: neurons are arranged in patterns, connecting with each other in changeable ways, depending on the conditions and requirements of the moment. Our photographable brain contains within it everything we need to create the mysteries of everything our invisible mind creates.

Notes

Chapter 1

1. L. Naci, R. Cusack, M. Anello, and A. M. Owen, "A Common Neural Code for Similar Conscious Experiences in Different Individuals," *Proceedings of the National Academy of Sciences* 111 (2014): 14277–14282.

2. This phrase is taken from the title of the biography *A Beautiful Mind*, by Sylvia Nasser (1998), about John Forbes Nash, who had schizophrenia, but whose brilliance earned him a Nobel Prize in economics.

3. F. C. Donders, "On the Speed of Mental Processes," in *Attention and Performance II: Acta Psychologica*, vol. 30, ed. W. G. Koster (1969), 412–431.(Original work published in 1868).

4. H. Ebbinghaus, *Memory: A Contribution to Experimental Psychology*, trans. H. A. Ruger and C. E. Bussenius (New York: Teachers College, Columbia University, 1913). (Original work, *Über das Gedächtnis*, published 1885.)

5. J. B. Watson, "Psychology as the Behaviorist Views It," *Psychological Review* 20 (1913): 158, 176; emphasis added.

6. J. B. Watson and R. Raynor, "Conditioned Emotional Reactions," *Journal of Experimental Psychology* 3 (1920): 1–14.

7. I. Pavlov, *Conditioned Reflexes* (London: Oxford University Press, 1927).

8. B. F. Skinner, *The Behavior of Organisms* (New York: Appleton Century, 1938).

9. F. J. Dyson, "Is Science Mostly Driven by Ideas or by Tools?" *Science* 338 (2012): 1426–1427; T. Kuhn, *The Structure of Scientific Revolutions* (Chicago: University of Chicago Press, 1962).

10. D. E. Broadbent, *Perception and Communication* (London: Pergamon Press, 1958).

11. E. C. Cherry, "Some Experiments on the Recognition of Speech, with One and with Two Ears," *Journal of the Acoustical Society of America* 25 (1953): 975–979.

12. J. McCarthy, M. L. Minsky, N. Rochester, and C. E. Shannon, "A Proposal for the Dartmouth Summer Research Project on Artificial Intelligence," August 31, 1955, http://www-formal.stanford.edu/jmc/history/dartmouth/dartmouth.html.

13. G. A. Miller, "The Magical Number Seven, Plus or Minus Two: Some Limits on Our Capacity for Processing Information," *Psychological Review* 63 (1956): 81–97.

14. W. Bechtel, A. Abrahamsen, and G. Graham, "The Life of Cognitive Science," in *A Companion to Cognitive Science*, ed. W. Bechtel and G. Graham, 2–104 (Oxford: Blackwell, 1998); G. A. Miller, "The Cognitive Revolution: A Historical Perspective," *Trends in Cognitive Sciences* 7 (2003): 141–144; U. Neisser, "New Vistas in the Study of Memory," in *Remembering Reconsidered: Ecological and Traditional Approaches to the Study of Memory*, ed. U. Neisser and E. Winograd, 1–10 (Cambridge: Cambridge University Press, 1988).

15. B. F. Skinner, *Verbal Behavior* (New York: Appleton-Century-Crofts, 1957).

16. N. Chomsky, "A Review of Skinner's *Verbal Behavior*," *Language* 35 (1959): 26–58.

17. U. Neisser, *Cognitive Psychology* (New York: Appleton-Century-Crofts, 1967).

18. Although behaviorism ceased to be the dominant paradigm in psychology, Skinner's accomplishment should not be diminished. His

explanation of how behavior can be controlled by reinforcements is still important, as evidenced by the everyday example of addictive cell phone usage, which is an example of reinforcements influencing behavior. The cognitive revolution pointed out that reinforcement theory not only ignores cognitive processes but also provides only a partial explanation of behavior.

19. A. Smith, *Powers of Mind* (New York: Simon & Schuster, 1975).

20. C. Castaneda, *Journey to Ixtlan: The Lessons of Don Juan* (New York: Simon & Schuster, 1972); A. Huxley, *The Doors of Perception* (London: Chatto & Windus, 1954).

21. J. R. MacLean, D. C. Macdonald, F. Oden, and E. Wilby, "LSD-25 and Mescaline as Therapeutic Adjuvants," in *The Use of LSD in Psychotherapy and Alcoholism*, ed. H. Abramson (New York: Bobbs-Merrill, 1967), 407–426.

22. M. Pollan, "Guided Explorations: My Adventures with the Researchers and Renegades Bringing Psychedelics into the Mental Health Mainstream," *New York Times Magazine*, May 20, 2018, 32–38, 61–65; M. Pollan, *How to Change Your Mind: What the New Science of Psychedelics Teaches Us about Consciousness, Dying, Addiction, Depression, and Transcendence* (New York: Penguin, 2018).

23. J. D. Creswell, "Mindfulness Interventions," *Annual Review of Psychology* 68 (2017): 491–516; J. P. Pozuelos, B. R. Mean, M. R. Rueda, and P. Malinowski, "Short-Term Mindful Breath Awareness Training Improves Inhibitory Control and Response Monitoring," *Progress in Brain Research* 244 (2019): 137–163.

24. R. Descartes, *Discourse on the Method* (1637).

25. Geshe Kelsang Gyatso, *Transform Your Life* (Ulverston, UK: Tharpa, 2002).

26. D. Chopra, "A Final Destination: The Human Universe," address to the Tucson Science of Consciousness Conference, April 27, 2016.

27. O. Blanke, T. Landis, L. Spinelli, and M. Seeck, "Out-of-Body Experience and Autoscopy of Neurological Origin," *Brain* 127 (2004): 243–258;

S. Bunning and O. Blanke, "The Out-of-Body Experience: Precipitating Factors and Neural Correlates," in *Progress in Brain Research*, vol. 150, ed. S. Laureys (New York: Elsevier, 2005).

28. O. Blanke and S. Arzy, "The Out-Of-Body Experience: Disturbed Self-Processing at the Temporo-Parietal Junction," *Neuroscientist* 11, no. 1 (2005): 16.

29. S. J. Blackmore, *Beyond the Body: An Investigation of Out-of-Body Experiences* (London: Heinemann, 1982); H. Irwin, *Flight of Mind: A Psychological Study of the Out-of-Body Experience* (Metuchen, NJ: Scarecrow Press, 1985).

30. D. De Ridder, K. Van Laere, P. Dupont, T. Monovsky, and P. Van de Hyning, "Visualizing Out-of-Body Experience in the Brain," *New England Journal of Medicine* 357 (2007): 1829–1833.

31. Bunning and Blanke, "The Out-of-Body Experience"; De Ridder et al., "Visualizing Out-of-Body Experience"; F. Tong, "Out-of-Body Experiences: From Penfield to Present," *Trends in Cognitive Sciences* 7 (2003): 104–106.

32. S. Blackmore, *Dying to Live: Near-Death Experiences* (New York: Prometheus Books, 1993); B. Greyson and I. Stevenson, "The Phenomenology of Near-Death Experiences," *American Journal of Psychiatry* 137 (1980): 1193–1196: K. R. Ring, *Life at Death: A Scientific Investigation of the Near-Death Experience* (New York: Coward, McCann & Geoghegan, 1980).

33. E. Alexander, *Proof of Heaven: A Neurosurgeon's Journey into the Afterlife* (New York: Simon & Schuster, 2013).

34. P. van Lommel, *Consciousness beyond Life: The Science of the Near-Death Experience* (New York: HarperCollins, 2010).

35. G. M. Woerlee, *Mortal Minds: The Biology of Near-Death Experience* (New York: Prometheus Books, 2005); G. M. Woerlee, "Review of P. M. Lommel, *Consciousness beyond Life*" (2019), http://neardth.com/consciousness-beyond-life.php.

36. S. Blackmore, *Consciousness*, 2nd ed. (New York: Routledge, 2010); L. Dittrich, "The Prophet," *Esquire*, July 2, 2013; O. Sacks, "Seeing God in the Third Millennium," *Atlantic*, December 12, 2012.

37. Sacks, "Seeing God in the Third Millennium."

38. R. L. Carhart-Harris, S. Muthukumaraswamy, L. Roseman, et al., Neural Correlates of the LSD Experience Revealed by Multimodal Neuroimaging," *Proceedings of the National Academy of Sciences* 113 (2016): 4853–4858.

39. J. Searle, "Theory of Mind and Darwin's Legacy," *Proceedings of the National Academy of Sciences* 110 (2013): 10343–10348.

40. J. Brockman, ed., *The Mind: Leading Scientists Explore the Brain, Memory, Personality, and Happiness* (New York: Harper, 2013).

41. D. S. Bassett and M. S. Gazzaniga, "Understanding Complexity in the Human Brain," *Trends in Cognitive Sciences* 15 (2011): 200–209.

42. P. Broca, "Sur le volume et la forme du cerveau suivant les individus et suivant les races," *Bulletin Societé d'Anthropologie Paris* 2 (1861): 139–207, 301–321, 441–446.

43. C. Wernicke, *Der aphasische Symptomenkomplex* (Breslau: Cohn, 1874).

44. Santiago Ramón y Cajal, "The Structure and Connections of Neurons: Nobel Lecture, December 12, 1906," in *Nobel Lectures, Physiology or Medicine, 1901–1921* (New York: Elsevier Science, 1967), 221–253.

45. E. R. Kandel, *In Search of Memory* (New York: Norton, 2006), 61.

46. E. D. Adrian, *The Basis of Sensation* (New York: Norton, 1928), 7; Adrian, *The Mechanism of Nervous Action* (Philadelphia: University of Pennsylvania Press, 1932).

47. Adrian, *The Basis of Sensation*, 7; *The Mechanism of Nervous Action*.

48. D. H. Hubel, "Exploration of the Primary Visual Cortex, 1955–1978," *Nature* 299 (1982): 515–524; D. H. Hubel and T. N. Wiesel, "Receptive Fields of Single Neurons in the Cat's Striate Cortex," *Journal of Physiology* 148 (1959): 574–591; D. H. Hubel and T. N. Wiesel, "Receptive Fields and Functional Architecture in Two Non-striate Visual Areas (18 and 19) of the Cat," *Journal of Neurophysiology* 28 (1965): 229–289.

49. H. Berger, *Psyche* (Jena: Gustav Fischer, 1940).

50. S. Ogawa, T. M. Lee, A. R. Kay, and D. W. Tank, "Brain Magnetic Resonance Imaging with Contrast Dependent on Blood Oxygenation," *Proceedings of the National Academy of Sciences*, 87 (1990): 9868–9872. M. M. Ter-Pogossian, M. E. Phelps, E. J. Hoffman, and N. A. Mullani, "A Positron-Emission Tomograph for Nuclear Imaging (PET)," *Radiology* 114 (1975): 89–98.

51. T. Kuhn, *The Structure of Scientific Revolutions* (Chicago: University of Chicago Press, 1962).

52. F. J. Dyson, "Is Science Mostly Driven by Ideas or by Tools?" *Science* 338 (2012): 1426–1427; P. Galison, *Image and Logic* (Chicago: University of Chicago Press, 1997).

53. P. T. Fox, "Human Brain Mapping: A Convergence of Disciplines," *Human Brain Mapping* 1 (1993): 1–2; A. Toga, "Editorial," *NeuroImage* 1 (1992): 1.

54. A. Eklund, T. E. Nichols, and H. Knutsson, "Cluster Failure: Why fMRI Inferences for Spatial Extent Have Inflated False-Positive Rates," *Proceedings of the National Academy of Sciences* 113 (2016): 7900–7905.

Chapter 2

1. The epigraph above comes from R. Van Gulick, "Consciousness," in *Stanford Encyclopedia of Philosophy* (Stanford, CA: Metaphysics Research Lab, 2014).

2. S. Blackmore, *Consciousness*, 2nd ed. (New York: Routledge, 2010).

3. J. Locke, *An Essay concerning Human Understanding* (1690).

4. D. Chalmers, "Facing Up to the Problem of Consciousness," *Journal of Consciousness Studies* 2, no. 3 (1995): 200–219.

5. Chalmers, "Facing Up."

6. Blackmore, *Consciousness*.

7. G. Smallberg, "No Shared Theory of Mind," in *What to Think about Machines That Think*, ed. J. Brockman (New York: Harper Perennial, 2015).

8. J. Searle, "Theory of Mind and Darwin's Legacy," *Proceedings of the National Academy of Sciences* 110, suppl. 2 (2013): 10343–10348.

9. D. Chalmers, "I'm Conscious: He's Just a Zombie," in *Conversations on Consciousness*, ed. S. Blackmore (New York: Oxford University Press, 2005).

10. B. Baars, "Consciousness Is a Real Working Theater," in *Conversations on Consciousness*, ed. S. Blackmore (New York: Oxford University Press, 2005).

11. W. James, *The Principles of Psychology*, rev. ed. (Cambridge, MA: Harvard University Press, 1981). (Original work published 1890.)

12. A. K. Seth, "The Grand Challenge of Consciousness," *Frontiers in Psychology* 1, no. 5 (2010): 1–2.

13. W. Seager, "Emergentist Panpsychism," *Journal of Consciousness Studies* 19, nos. 9–10 (2012): 19–39; N. D. Theise and M. C. Kafatos, "Complementarity in Biological Systems: A Complexity View," *Complexity* 18, no. 6 (2013): 11–20; G. Tononi and C. Koch, "Consciousness: Here, There and Everywhere?" *Philosophical Transactions of the Royal Society B* 370 (2015): 20140167.

14. C. McGinn, *The Mysterious Flame: Conscious Minds in a Material World* (New York: Basic Books, 1999); J. Searle, "Consciousness and the Philosophers," *New York Review of Books* 44, no. 4 (1997): 43–50.

15. P. Tompkins and C. Bird, *The Secret Life of Plants* (New York: Harper & Row, 1973).

16. M. Pollan, "The Intelligent Plant," *New Yorker*, December 23, 30, 2013.

17. D. Chamovitz, *What a Plant Knows* (New York: Scientific American/Farrar, Straus and Giroux, 2012).

18. Another book that is scientific but draws parallels between trees and animals in a way similar to Chamovitz's book on plants is P. Wohlleben, *The Hidden Life of Trees: What They Feel, How They Communicate* (Greystone Books, 2016). See also K. Yokawa et al., "Anaesthetics Stop

Diverse Plant Organ Movements, Affect Endocytic Vesicle Recycling and ROS Homeostasis, and Block Action Potentials in Venus Flytraps," *Annals of Botany* 122 (2018): 747–756. This paper shows that anesthetics affect plants and animals in similar ways.

19. But if you want to be treated to perhaps the best example of anthropomorphism available in novel form, see Jack London, *Call of the Wild* (1903), which is written from the point of view of a dog named Buck.

20. J. Riley, U. Greggers, A. Smith, D. Reynolds, and R. Menzeil, "The Flight Paths of Honeybees Recruited by the Waggle Dance," *Nature* 435, no. 7039 (2005): 205–207.

21. M. Boly, A. K. Seth, M. Wilke, P. Ingmundson, B. Baars, S. Laureys, D. B. Edelman, and N. Tsuchiya, "Consciousness in Humans and Nonhuman animals: Recent Advances and Future Directions," *Frontiers in Psychology* 4, no. 625 (2013): 1–20.

22. D. Bilefsky, "Inky the Octopus Escapes from a New Zealand Aquarium," *New York Times*, April 14, 2016.

23. J. A. Mather, "Cephalopod Consciousness: Behavioral Evidence," *Consciousness and Cognition* 17 (2008): 37–48.

24. C. B. Albertin, O. Simakov, T. Mitros, Z. Yan Wang, J. R. Pungor, E. Edsinger-Gonzales, C. Brenner, C. W. Ragsdale, and D. S. Rokhsar, "The Octopus Genome and the Evolution of Cephalopod Neural and Morphological Novelties," *Nature* 524 (2015): 220–224.

25. A. Abbott, "DNA Sequence Expanded in Areas Otherwise Reserved for Vertebrates," *Nature News*, August 12, 2015.

26. A. B. Barron and C. Klein, "What Insects Can Tell Us about the Origins of Consciousness," *Proceedings of the National Academy of Sciences B* 113, no. 18 (2016): 4900–4908.

27. Barron and Klein, "What Insects Can Tell Us."

28. A. Damasio, H. Damasio, and D. Travel, "Persistence of Feelings and Sentience after Bilateral Damage of the Insula," *Cerebral Cortex* 23, no. 4 (2013): 833–846; B. Merker, "Consciousness without a Cerebral

Cortex: A Challenge for Neuroscience and Medicine," *Behavioral Brain Sciences* 30, no. 1 (2007): 63–81, discussion at 81–134; A. M. Owen et al., "Detecting Residual Cognitive Function in Persistent Vegetative State," *Neurocase* 8, no. 5 (2002): 395–403.

29. Barron and Klein, "What Insects Can Tell Us."

30. Damasio, Damasio, and Travel, "Persistence of Feelings and Sentience"; Merker, "Consciousness without a Cerebral Cortex"; Owen et al., "Detecting Residual Cognitive Function."

31. A. Avargues-Weber and M. Giurfa, "Conceptual Learning by Miniature Brains," *Proceedings of the Royal Society B* 280 (2013): 1–9; D. Van Essen, "Organization of visual areas in macaque and human cerebral cortex," in *The Visual Neurosciences*, ed. L. M. Chalupa and J. S. Werner (Cambridge, MA: MIT Press, 2004), 507–521.

32. D. B. Edelman, B. J. Baars, and A. K. Seth, "Identifying Hallmarks of Consciousness in Non-mammalian Species," *Consciousness and Cognition* 14 (2005): 169–187; J. Panksepp, "Affective Consciousness: Core Emotional Feelings in Animals and Humans," *Consciousness and Cognition* 14 (2005): 30–80.

33. "The Cambridge Declaration on Consciousness," Francis Crick Memorial Conference on Consciousness in Human and Non-Human Animals, Churchill College, University of Cambridge, July 7, 2012, http://fcmconference.org/img/CambridgeDeclarationOnConsciousness.pdf.

34. Chalmers, "I'm Conscious: He's Just a Zombie."

35. L. Mudrik, N. Faivre, and C. Koch, "Information Integration without Awareness," *Trends in Cognitive Sciences* 18 (2014): 488–496; A. K. Seth, E. Izhikevich, G. N. Reeke, and G. M. Edelman, "Theories and Measures of Consciousness: An Extended Framework," *Proceedings of the National Academy of Sciences* 103 (2006): 10799–10804; G. Tononi and C. Koch, "The Neural Correlates of Consciousness: An Update," *Annals of the New York Academy of Sciences* 1124 (2008): 239–261; G. Tononi and G. M. Edelman, "Consciousness and Complexity," *Science* 282 (1998): 1846–1851.

36. T. Nagel, "What Is It Like to Be a Bat?" *Philosophical Review* 83, no. 4 (1974): 436.

37. G. Tononi and C. Koch, "Consciousness: Here, There and Everywhere?" *Philosophical Transactions of the Royal Society B* 370 (2015): 20140167.

38. D. Chalmers, *The Conscious Mind* (New York: Oxford University Press, 1996).

39. F. Jackson, "What Mary Didn't Know," *Journal of Philosophy* 83 (1986): 291–295.

40. E. B. Goldstein and J. Brockmole, *Sensation and Perception*, 10th ed. (Boston: Cengage, 2017); see chap. 10 for a discussion of binocular vision.

41. S. Barry, *Fixing My Gaze* (New York: Basic Books, 2009).

42. I. Newton, *Optiks* (London: Smith and Walford, 1704).

43. R. Menzel and W. Backhaus, "Color Vision in Honeybees: Phenomena and Physiological Mechanisms," in *Facets of Vision*, ed. D. G. Stavenga and R. C. Hardie (Berlin: Springer, 1989), 281–297; R. Menzel, D. F. Ventura, H. Hertel, J. deSouza, and U. Greggers, "Spectral Sensitivity of Photoreceptors in Insect Compound Eyes: Comparison of Species and Methods," *Journal of Comparative Physiology* 158A (1986): 165–177.

44. Chalmers, "Facing Up," 220; D. Chalmers, "Moving Forward on the Problem of Consciousness," *Journal of Consciousness Studies* 4, no. 1 (1997): 3–46.

45. J. Brockman, ed., *The Mind* (New York: Harper, 2011).

46. V. J. Kafalov, "Rod and Cone Visual Pigments and Phototransduction through Pharmacological, Genetic, and Physiological Approaches," *Journal of Biological Chemistry* 287, no. 3 (2012): 1635–1641.

47. Brockman, *The Mind*.

48. Chalmers, "Facing Up," 220; Chalmers, "Moving Forward."

49. T. H. Huxley, *Evidence as to Men's Place in Nature* (London: Williams and Wilkins, 1863).

50. J. Levine, "Materialism and Qualia: The Explanatory Gap," *Pacific Philosophical Quarterly* 64 (1983): 354–361.

51. W. T. Newsome, K. H. Britten, and J. A. Movshon, "Neuronal Correlates of a Perceptual Decision," *Nature* 341 (1989): 52–54.

52. A. Gopnik, "Amazing Babies," in *The Mind*, ed. J. Brockman (New York: Harper Perennial, 2011).

53. D. E. Stansbury, T. Naselaris, and J. L. Gallant, "Natural Scene Statistics Account for the Representation of Scene Categories in Human Visual Cortex," *Neuron* 79 (2013): 1025–1034.

54. A. G. Huth, S. Nishimoto, A. T. Vo, and J. L. Gallant, "A Continuous Semantic Space Describes the Representation of Thousands of Objects and Action Categories across the Human Brain," *Neuron* 76 (2012): 1210–1224; A. G. Huth, W. A. deHeer, T. L. Griffiths, F. E. Theunissen, and J. L. Gallant, "Natural Speech Reveals the Semantic Maps That Tile Human Cerebral Cortex," *Nature* 532 (2016): 453–460; T. Naselaris, C. A. Olman, D. E. Stansburh, K. Uurbil, and J. Gallant, "A Voxel-Wise Encoding Model for Early Visual Areas Decodes Mental Images of Remembered Scenes," *NeuroImage* 105 (2015): 215–228.

55. D. S. Bassett and M. S. Gazzaniga, "Understanding Complexity in the Human Brain," *Trends in Cognitive Sciences* 15 (2011): 200–209.

56. S. E. Henschen, *Klinische und anatomische Beitrage zue Pathologie des Gehirns (Pt. 1)* (Almquist and Wiksell, 1890); H. Munk, *Uber die Funktionen der Grosshirnrinde* (A. Hirschwald, 1881) (English translation in *The Cerebral Cortex*, ed. G. Van Bonin [Springfield: Thomas, 1960], 97–117).

57. S. Corkin, "What's New with the Amnesic Patient H.M.?" *Nature Reviews Neuroscience* 3 (2002): 1–8; W. B. Scoville and B. Milner, "Loss of Recent Memory after Bilateral Hippocampus Lesions," *Journal of Neurosurgery and Psychiatry* 20 (1957): 11–21.

58. D. Tranel, G. Gullickson, M. Koch, and R. Adolphs, "Altered Experience of Emotion following Bilateral Amygdala Damage," *Cognitive Neuropsychiatry* 11 (2007): 219–232.

59. R. Adolphs, J. A. Russell, and D. Travel, "A Role for the Human Amygdala in Recognizing Emotional Arousal from Unpleasant Stimuli," *Psychological Science* 10 (1999): 167–171.

60. J. S. Morris, C. D. Frith, D. I. Perrett, A. W. Rowland, A. J. Calder, and R. J. Dolan, "A Differential Neural Response in the Human Amygdala to Fearful and Happy Facial Expressions," *Nature* 383 (1996): 812–815.

61. D. H. Hubel, "Exploration of the Primary Visual Cortex, 1955–1978," *Nature* 299 (1982): 515–524; D. H. Hubel and T. N. Wiesel, "Receptive Fields of Single Neurons in the Cat's Striate Cortex," *Journal of Physiology* 148 (1959): 574–591; D. H. Hubel and T. N. Wiesel, "Receptive Fields and Functional Architecture in Two Non-striate Visual Areas (18 and 19) of the Cat," *Journal of Neurophysiology* 28 (1965): 229–289.

62. P. E. Downing, J. Yuhong, M. Shuman, and N. Kanwisher, "A Cortical Area Selective for Visual Processing of the Human Body," *Science* 239 (2001): 2470–2473; R. Epstein, A. Harris, D. Stanley, and N. Kanwisher, "The Parahippocampal Place Area: Recognition, Navigation, or Encoding?" *Neuron* 23 (2001): 115–125; C. G. Gross, C. E. Rocha-Miranda, and D. B. Bender, "Visual Properties of Neurons in Inferotemporal Cortex of the Macaque," *Journal of Neurophysiology* 5 (1972): 96–111; N. Kanwisher, "The Ventral Visual Object Pathway in Humans: Evidence from fMRI," in *The Visual Neurosciences*, ed. L. M. Chalupa and J. S. Werner (Cambridge, MA: MIT Press, 2003), 1179–1190; N. Kanwisher and D. D. Dilks, "The Functional Organization of the Ventral Visual Pathway in Humans," in *The New Visual Neurosciences*, ed. J. S. Werner and L. M. Chalupa (Cambridge, MA: MIT Press, 2013); N. Kanwisher, J. McDermott, and M. M. Chun, "The Fusiform Face Area: A Module in Human Extrastriate Cortex Specialized for Face Perception," *Journal of Neuroscience* 17 (1997): 4302–4311; D. I. Perrett, E. T. Rolls, and W. Caan, "Visual Neurons Responsive to Faces in the Monkey Temporal Cortex," *Experimental Brain Research* 7 (1982): 329–342; E. T. Rolls, "Responses of Amygdaloid Neurons in the Primate," in *The Amygdaloid Complex*, ed. Y. Ben-Ari (Amsterdam: Elsevier, 1981), 383–393.

63. D. C. Van Essen, "Corticocortical and Thalamocortical Information Flow in the Primate Visual System," *Progress in Brain Research* 149 (2005): 173–185.

64. M. Behrmann and D. C. Plaut, "Distributed Circuits, Not Circumscribed Centers, Mediate Visual Recognition," *Trends in Cognitive Sciences* 17 (2013): 210–219.

65. Huth et al., "Continuous Semantic Space"; E. K. Warrington and T. Shallice, "Category Specific Semantic Impairments," *Brain* 107 (1984): 829–854.

66. C. E. Curtis and M. D'Esposito, "Persistent Activity in the Prefrontal Cortex during Working Memory," *Trends in Cognitive Sciences* 7 (2003): 415–423; S. A. Harrison and F. Tong, "Decoding Reveals the Contents of Visual Working Memory in Early Visual Areas," *Nature* 458 (2009): 462–465.

67. B. Levine, G. R. Turner, D. Tisserand, S. J. Hevenor, S. J. Graham, and A. R. McIntosh, "The Functional Neuroanatomy of Episodic and Semantic Autobiographical Remembering: A Prospective Functional MRI Study," *Journal of Cognitive Neuroscience* 16 (2004): 1633–1646.

Chapter 3

1. O. Sacks, *The Man Who Mistook His Wife for a Hat* (New York: Simon & Schuster, 1985).

2. C. S. Konen, M. Behrmann, M. Nishimura, and S. Kastner, "The Functional Neuroanatomy of Object Agnosia: A Case Study," *Neuron* 71 (2011): 49–60.

3. A. D. Milner, D. I. Perrett, R. S. Johnston, P. J. Benson, T. R. Jordan, D. W. Heeley, D. Bettucci, F. Mortara, R. Mutani, E. Terrazzi, and D. L. W. Davidson, "Perception and Action in 'Visual Form Agnosia,'" *Brain* 114 (1991): 405–428; A. D. Milner and M. A. Goodale, "The Visual Brain in Action," *Psyche* 4, no. 12 (1998).

4. Milner et al., "Perception and Action in 'Visual Form Agnosia.'" Fig. 3.1 is based on fig. 2 in Milner and Goodale, "The Visual Brain in Action."

5. M. A. Goodale, "How (and Why) the Visual Control of Action Differs from Visual Perception," *Proceedings of the Royal Society of London B* 281 (2014): 20140337; M. A. Goodale, D. A. Westwood, and A. D. Milner,

"Two Distinct Modes of Control for Object-Directed Action," *Progress in Brain Research* 144 (2004): 131–144.

6. M. Mishkin, L. G. Ungerleider, and K. G. Macko, "Object Vision and Spatial Vision: Two Cortical Pathways," *Trends in Neuroscience* 6 (1983): 414–417; L. G. Ungerleider and M. Mishkin, "Two Cortical Visual Systems," in *Analysis of Visual Behavior*, ed. D. J. Ingle, M. A. Goodale, and R. J. W. Mansfield (Cambridge, MA: MIT Press, 1982), 549–586. Fig. 3.2 is from E. B. Goldstein and J. Brockmole, *Sensation and Perception*, 10th ed. (Boston: Cengage, 2017), fig. 414, 80; adapted from Mishkin, Ungerleider, and Macko, "Object Vision and Spatial Vision."

7. Mishkin, Ungerleider, and Macko, "Object Vision and Spatial Vision"; Ungerleider and Mishkin, "Two Cortical Visual Systems."

8. F. Fang and S. He, "Cortical Responses to Invisible Objects in the Human Dorsal and Ventral Pathways," *Nature Neuroscience* 8, no. 10 (2005): 1380–1385; Goodale, "How (and Why) the Visual Control of Action Differs from Visual Perception."

9. V. S. Ramachandran and S. Blakeslee, *Phantoms in the Brain: Probing the Mysteries of the Human Mind* (New York: William Morrow, 1988), 64–65.

10. G. Riddoch, "Dissociation of Visual Perceptions due to Occipital Injuries, with Especial Reference to Appreciation of Movement," *Brain* 40 (1917): 15–57.

11. L. Weiskrantz, E. K. Warrington, M. D. Sanders, and J. Marshall, "Visual Capacity in the Hemianopic Field Following a Restricted Occipital Ablation," *Brain* 97 (1974): 709–728.

12. V. A. F. Lamme, "Blindsight: The Role of Feedforward and Feedback Corticocortical Connections," *Acta Psychologica* 107 (2001): 209–228.

13. J. Driver and P. Vuilleumier, "Perceptual Awareness and Its Loss in Unilateral Neglect and Extinction," *Cognition* 79 (2001): 39–88.

14. Fig. 3.3 is adapted from fig. 1b in P. Vuilleumier and S. Schwartz, "Emotional Facial Expressions Capture Attention," *Neurology* 56 (2001): 153–158.

15. P. Vuilleumier and S. Schwartz, "Beware and Be Aware: Capture of Spatial Attention by Fear-Related Stimuli in Neglect," *NeuroReport* 12, no. 6 (2001): 1119–1122.

16. Vuilleumier and Schwartz, "Emotional Facial Expressions Capture Attention," "Beware and Be Aware"; R. D. Rafel, "Neglect," *Current Opinion in Neurobiology* 4 (1994): 231–236.

17. A. Dijksterhuis and H. Aarts, "Goals, Attention, and (Un)consciousness," *Annual Review of Psychology* 61 (2010): 467–490; B. Libet, C. A. Gleason, E. W. Wright, and D. K. Pearl, "Time of Conscious Intention to Act in Relation to Onset of Cerebral Activity (Readiness-Potential)," *Brain* 106 (1983): 623–642.

18. S. Bode, A. H. He, C. S. Soon, R. Trampel, R. Turner, and J.-D. Haynes, "Tracking the Unconscious Generation of Free Decisions Using Ultra-High Field fMRI," *PLoS One* 6, no. 6 (2011): e21612; I. Fried, R. Mukamel, and G. Kreiman, "Internally Generated Preactivation of Single Neurons in Human Medial Frontal Cortex Predicts Volition," *Neuron* 69 (2011): 548–562; P. Haggard, "Human Volition: Towards a Neuroscience of Will," *Nature Reviews Neuroscience* 9 (2008): 934–946; C. S. Soon, M. Brass, H.-J. Heinze, and J.-D. Haynes, "Unconscious Determinants of Free Decisions in the Human Brain," *Nature Neuroscience* 11, no. 5 (2008): 543–545.

19. Fried, Mukamel, and Kreiman, "Internally Generated Preactivation of Single Neurons."

20. Soon et al., "Unconscious Determinants of Free Decisions," 543.

21. Y. H. R. Kang, F. H. Petzschner, D. M. Wolpert, and M. N. Shadlen, "Piercing of Consciousness as a Threshold-Crossing Operation," *Current Biology* 27 (2017): 2285–2295; J. Miller and W. Schwarz, "Brain Signals Do Not Demonstrate Unconscious Decision Making: An Interpretation Based on Graded Conscious Awareness," *Consciousness and Cognition* 24 (2014): 12–21.

22. Fig. 10 from Miller and Schwarz, "Brain Signals Do Not Demonstrate."

23. Kang et al., "Piercing of Consciousness"; Miller and Schwarz, "Brain Signals Do Not Demonstrate."

24. Soon et al., "Unconscious Determinants of Free Decisions," 543.

25. P. Haggard, "Human Volition: Towards a Neuroscience of Will," *Nature Reviews Neuroscience* 9 (2008): 942.

26. S. Joordens, M. van Duijn, and T. M. Spalek, "When Timing the Mind One Should Also Mind the Timing: Biases in the Measurement of Voluntary Actions," *Consciousness and Cognition* 11 (2002): 231–240.

27. P. Alexander, S. Alexander, W. Sinnott-Armstrong, A. L. Roskies, T. Wheatley, and P. U. Tse, "Readiness Potentials Driven by Non-motoric Processes," *Consciousness and Cognition* 39 (2016): 38–47; H.-G. Jo, T. Hinterberger, M. Wittmann, T. L. Borghardt, and S. Schmidt, "Spontaneous EEG Fluctuations Determine the Readiness Potential: Is Preconscious Brain Activation a Preparation Process to Move?" *Experimental Brain Research* 231 (2013): 495–500.

28. B. Libet, "Unconscious Cerebral Initiative and the Role of Conscious Will in Voluntary Action," *Behavioral and Brain Sciences* 8 (1985): 529–566. Libet's paper is on 529–539; open peer commentary on the paper, plus Libet's response, is on 539–566.

29. A. S. Reber, "Implicit Learning of Artificial Grammars," *Journal of Verbal Learning and Verbal Behavior* 6 (1967): 855–863.

30. Reber, "Implicit Learning of Artificial Grammars."

31. J. R. Saffran, R. Aslin, and E. L. Newport, "Statistical Learning by 8-Month-Old Infants," *Science* 274 (1996): 1926–1928.

32. D. L. Schacter and R. L. Buckner, "Priming and the Brain," *Neuron* 20 (1998): 185–195.

33. Schacter and Buckner, "Priming and the Brain."

34. A. C. Kay, S. C. Wheeler, J. A. Bargh, and L. Ross, "Material Priming: The Influence of Mundane Physical Objects on Situational Construal and Competitive Behavioral Choice," *Organizational Behavior and Human Decision Processes* 95 (2004): 83–96.

35. M. Bateson, D. Nettle, and G. Roberts, "Cues of Being Watched Enhance Cooperation in a Real-World Setting," *Biological Letters* 2 (2006): 412–414.

36. R. W. Holland, M. Hendricks, and H. Aarts, "Smells like Clean Spirit," *Psychological Science* 16, no. 9 (2005): 689–693.

Chapter 4

1. A. Charpentier, "Analyze experimentale quelques de la sensation de poids" [Experimental study of some aspects of weight perception], *Archives de Physiologie Normales et Pathologiques* 3 (1891): 122–135.

2. G. Buckingham, "Getting a Grip on Heaviness Perception: A Review of Weight Illusions and Their Possible Causes," *Experimental Brain Research* 232 (2014): 1623–1629.

3. A. Clark, "Whatever Next? Predictive Brains, Situated Agents, and the Future of Cognitive Science," *Behavioral and Brain Sciences* 36 (2013): 181–253.

4. P. Kok, G. J. Brouwer, M. A. J. van Gerven, and F. P. de Lange, "Prior Expectations Bias Sensory Representation in Visual Cortex," *Journal of Neuroscience* 33, no. 41 (2013): 16275.

5. H. Helmholtz, *Handbuch der physiologischen optic*, vol. 3 (1860); English translation by J. P. C. Southall (New York: Dover, 1962); R. M. Warren R. P. Warren, *Helmholtz on Perception, Its Physiology and Development* (New York: Wiley, 1968).

6. Fig. 4.1 is adapted from fig. 3.7 in E. B. Goldstein, *Cognitive Psychology*, 5th ed. (San Francisco: Cengage, 2019), 65.

7. Fig. 4.2 is adapted from fig. 3.14 in Goldstein, *Cognitive Psychology*, 70.

8. Helmholtz, *Handbuch der physiologischen optic*, vol. 3; Warren and Warren, *Helmholtz on Perception*.

9. H. K. Hartline, "The Receptive Fields of Optic Nerve Fibers," *American Journal of Physiology* 130 (1940): 690–699.

10. D. H. Hubel and T. N. Wiesel, "Receptive Fields of Single Neurons in the Cat's Striate Cortex," *Journal of Physiology* 148 (1959): 574–591; D. H. Hubel and T. N. Wiesel, "Integrative Action in the Cat's Lateral

Geniculate Body," *Journal of Physiology* 155 (1961): 385–398; D. H. Hubel and T. N. Wiesel, "Receptive Fields and Functional Architecture in Two Non-striate Visual Areas (18 and 19) of the Cat," *Journal of Neurophysiology* 28 (1965): 229–289.

11. C. G. Gross, C. E. Rocha-Miranda, and D. B. Bender, "Visual Properties of Neurons in Inferotemporal Cortex of the Macaque," *Journal of Neurophysiology* 5 (1972): 96–111; N. Kanwisher, "The Ventral Visual Object Pathway in Humans: Evidence from fMRI," in *The Visual Neurosciences*, ed. L. M. Chalupa and J. S. Werner (Cambridge, MA: MIT Press, 2003), 1179–1190; N. Kanwisher, J. McDermott, and M. M. Chun, "The Fusiform Face Area: A Module in Human Extrastriate Cortex Specialized for Face Perception," *Journal of Neuroscience* 17 (1997): 4302–4311; E. T. Rolls and M. J. Tovee, "Sparseness of the Neuronal Representation of Stimuli in the Primate Temporal Visual Cortex," *Journal of Neurophysiology* 73 (1995): 713–726.

12. R. L. Gregory, *Eye and Brain* (New York: McGraw Hill, 1966).

13. R. L. Gregory, *Eye and Brain*, 3rd ed. (New York: McGraw Hill, 1978).

14. R. L. Gregory, *Eye and Brain*, 5th ed. (New York: McGraw Hill, 1997).

15. J. Anderson, H. B. Barlow, and R. L. Gregory, "Introduction to 'Knowledge-Based Vision in Man and Machine': A Discussion Held at the Royal Society," *Philosophical Transactions of the Royal Society of London B* 352 (1997): 1117–1120; M. Bar, "Predictions: A Universal Principle in the Operation of the Human Brain: Introduction to Theme Issue 'Prediction in the Brain: Using Our Past to Prepare for the Future,'" *Philosophical Transactions of the Royal Society of London B* 364 (2009): 1181–1182.

16. M. Bar, "The Proactive Brain: Memory for Predictions," *Philosophical Transactions of the Royal Society B* 364 (2009): 1235–1243; A. Bubic, D. Y. von Cramon, and R. I. Schubotz, "Prediction, Cognition and the Brain," *Frontiers in Human Neuroscience* 4 (2010): article 25; J. Hohwy, *The Predictive Mind* (New York: Oxford, 2013).

17. A. Oliva and A. Torralba, "The Role of Context in Object Recognition," *Trends in Cognitive Sciences* 11 (2007): 520–527.

18. Fig. 4.3 is adapted from Oliva and Torralba, "Role of Context."

19. Goldstein, *Cognitive Psychology*, 5th ed.

20. D. A. Kleffner and V. S. Ramachandran, "On the Perception of Shape from Shading," *Perception and Psychophysics* 52 (1992): 18–36.

21. Fig. 4.4 is from Goldstein, *Cognitive Psychology*, 5th ed., 75.

22. L. F. Barrett and M. Bar, "See It with Feeling: Affective Predictions during Object Perception," *Philosophical Transactions of the Royal Society B* 364 (2009): 1325.

23. T. Bayes, "An Essay towards Solving a Problem in the Doctrine of Chances," *Philosophical Transactions of the Royal Society of London* 53 (1763): 370–418.

24. K. P. Körding and D. M. Wolpert, "Bayesian Decision Theory in Sensorimotor Control," *Trends in Cognitive Science* 10 (2006): 319–326; J. B. Tenenbaum, C. Kemp, T. L. Griffiths, and N. D. Goodman, "How to Grow a Mind: Statistics, Structure, and Abstraction," *Science* 331 (2011): 1279–1285.

25. A. I. Dale, *A History of Inverse Probability from Thomas Bayes to Karl Pearson*, 2nd ed. (New York: Springer, 1999); G. Westheimer, "Was Helmholtz a Bayesian?" *Perception* 37 (2008): 642–650.

26. D. Kersten, P. Mamassian, and A. Yuille, "Object Perception as Bayesian Inference," *Annual Review of Psychology* 55 (2004): 271–304; D. C. Knill and A. Pouget, "The Bayesian Brain: The Role of Uncertainty in Neural Coding and Computation," *Trends in Neurosciences* 27 (2004): 712–719; D. C. Knill and W. Richards, *Perception and Bayesian Inference* (Cambridge: Cambridge University Press, 1996).

27. Kersten, Mamassian, and Yuille, "Object Perception as Bayesian Inference."

28. J. S. Bowers and C. J. Davis, "Bayesian Just-So Stories in Psychology and Neuroscience," *Psychological Bulletin* 138 (2012): 389–414.

29. M. Bar, "The Proactive Brain: Using Analogies and Associations to Generate Predictions," *Trends in Cognitive Sciences* 11 (2007): 280–289;

P. Kok, G. J. Brouwer, M. A. J. van Gerven, and F. P. de Lange, "Prior Expectations Bias Sensory Representations in Visual Cortex," *Journal of Neuroscience* 33 (2013): 16275–16284; A. Pouget, J. M. Beck, W. J. Ma, and P. E. Latham, "Probabilistic Brains: Knowns and Unknowns," *Nature Neuroscience* 16 (2013): 1170–1178.

30. M. Bar, K. S. Kassam, A. S. Ghuman, J. Boshyan, A. M. Schmid, A. M. Dale, M. S. Hamalainen, K. Marinkovic, D. L. Schacter, B. R. Rosen, and E. Halgren, "Top-Down Facilitation of Visual Recognition," *Proceedings of the National Academy of Sciences* 103 (2006): 449–454.

31. W. Schultz, P. Dayan, and P. R. Montague, "A Neural Substrate of Prediction and Reward," *Science* 275 (1997): 1593–1599.

32. T. Meyer and C. R. Olson, "Statistical Learning of Visual Transitions in Monkey Inferotemporal Cortex," *Proceedings of the National Academy of Sciences* 108 (2011): 19401–19406; T. Meyer, S. Ramachandran, and C. R. Olson, "Statistical Learning of Serial Visual Transitions by Neurons in Monkey Inferotemporal Cortex," *Journal of Neuroscience* 34 (2014): 9332–9337.

33. A. Alink, C. M. Schwiedrzik, A. Kohler, W. Singer, and L. Muckli, "Stimulus Predictability Reduces Responses in Primary Visual Cortex," *Journal of Neuroscience* 30 (2010): 2960–2966; A. Brodski, G.-F. Paasch, S. Helbling, and M. Wibral, "The Faces of Predictive Coding," *Journal of Neuroscience* 35 (2015): 8997–9006.

34. H. E. M. den Ouden, P. Kok, and F. P. de Lange, "How Prediction Errors Shape Perception, Attention, and Motivation," *Frontiers of Psychology* 3 (2012): article 548; M. F. Panichello, O. S. Cheung, and M. Bar, "Predictive Feedback and Conscious Visual Experience," *Frontiers in Psychology* 3 (2013): article 620.

35. Clark, "Whatever Next?"; J. Koster-Hale and R. Saxe, "Theory of Mind: A Neural Prediction Problem," *Neuron* 79 (2013): 836–838.

36. R. P. N. Rao and D. H. Ballard, "Predictive Coding in the Visual Cortex: A Functional Interpretation of Some Extra-Classical Receptive Field Effects," *Nature Neuroscience* 2 (1999): 79–87; A. K. Seth,

"Interoceptive Inference, Emotion, and the Embodied Self," *Trends in Cognitive Sciences* 17 (2013): 565–573.

37. Barrett and Bar, "See It with Feeling"; K. Friston and S. Kiebel, "Predictive Coding under the Free-Energy Principle," *Transactions of the Royal Society B* 364 (2009): 1211–1221.

38. A. Clark, "Whatever Next?"; R. P. N. Rao and D. H. Ballard, "Predictive Coding in the Visual Cortex: A Functional Interpretation of Some Extra-Classical Receptive Field Effects," *Nature Neuroscience* 2 (1999): 79–87.

39. A. Clark, *Surfing Uncertainty* (New York: Oxford University Press, 2016).

40. Fig. 4.6 is adapted from fig. 6.2 in E. B. Goldstein, *Sensation and Perception*, 9th ed. (Boston: Cengage, 2014); Courtesy of John Henderson.

41. J. M. Henderson, "Gaze Control as Prediction," *Trends in Cognitive Sciences* 21 (2017): 15–23.

42. J. Jovancevic-Misic and M. Hayhoe, "Adaptive Gaze Control in Natural Environments," *Journal of Neuroscience* 29 (2009): 6234–6238; Henderson, "Gaze Control as Prediction."

43. M. L. H. Võ and J. M. Henderson, "Does Gravity Matter? Effects of Semantic and Syntactic Inconsistencies on the Allocation of Attention during Scene Perception," *Journal of Vision* 9 (2009): 24.1–24.15.

44. M. Hayhoe and D. Ballard, "Eye Movements in Natural Behavior," *Trends in Cognitive Sciences* 9 (2005): 188–193; M. F. Land and M. Hayhoe, "In What Ways Do Eye Movements Contribute to Everyday Activities?" *Vision Research* 41 (2001): 3559–3565; B. W. Tatler, M. M. Hayhoe, M. F. Land, and D. H. Ballard, "Eye Guidance in Natural Vision: Reinterpreting Salience," *Journal of Vision* 11 (2011): 1–23.

45. R. W. Sperry, "Neural Basis of the Spontaneous Optokinetic Response Produced by Visual Inversion," *Journal of Comparative and Physiological Psychology* 43 (1950): 482–489; E. von Holst and H. Mittelstaedt, "Das Reafferenzprinzip: Wechselwirkungen zwischen Zentralnervensystem und Peripherie," *Naturwissenschaften* 37 (1950): 464–476.

46. M. A. Sommer and R. H. Wurtz, "A Pathway in Primate Brain for Internal Monitoring of Movements," *Science* 296 (2002): 1480–1482; M. A. Sommer and R. H. Wurtz, "Brain Circuits for the Internal Monitoring of Movements," *Annual Review of Neuroscience* 31 (2008): 317–338; R. W. Wurtz, "Corollary Discharge in Primate Vision," *Scholarpedia* 8, no. 10 (2013): 12335. These references describe the corollary discharge process in more detail than the simplified diagram in figure 4.7. One feature of the more complex explanation is a number of brain structures, which are collectively called the "comparator," that receive both the visual signal (or auditory signal for hearing or somatosensory for touch) and the corollary discharge signal. The signal from the comparator then influences perception.

47. R. W. Wurtz, K. McAlonan, J. Cavanaugh, and R. A. Berman, "Thalamic Pathways for Active Vision," *Trends in Cognitive Sciences* 15 (2011): 177–184; Wurtz, "Corollary Discharge in Primate Vision."

48. J. F. A. Poulet and B. Hedwig, "A Corollary Discharge Maintains Auditory Sensitivity during Sound Production," *Nature* 418 (2002): 872–876; J. F. A. Poulet and B. Hedwig, "Corollary Discharge Inhibition of Ascending Auditory Neurons in the Stridulating Cricket," *Journal of Neuroscience* 23 (2003): 4717–4725.

49. P. M. Bays, J. R. Flanagan, and D. M. Wolpert, "Attenuation of Self-Generated Tactile Sensations Is Predictive, Not Postdictive," *PLoS Biology* 4, no. 2 (2006): 281–284.

50. S.-J. Blakemore, C. D. Frith, and D. M. Wolpert, "Spatio-temporal Prediction Modulates the Perception of Self-Produced Stimuli," *Journal of Cognitive Neuroscience* 11 (1999): 551–559.

51. S.-J. Blakemore, D. M. Wolpert, and C. D. Frith, "Central Cancellation of Self-Produced Tickle Sensation," *Nature Neuroscience* 1 (1998): 635–640.

52. D. M. Wolpert and J. R. Flanagan, "Motor Prediction," *Current Biology* 11 (2001): R729–R732.

53. H. Scherberger, R. Q. Quiroga, and R. A. Anderson, "Coding of Movement Intentions," in *Principles of Neural Coding*, ed. R. Q. Quiroga and S. Panzeri (Taylor & Francis, 2013), 303–321.

54. P. Fattori, V. Raos, R. Breveglieri, A. Bosco, N. Marzocchi, and C. Galletti, "The Dorsomedial Pathway Is Not Just for Reaching: Grasping Neurons in the Medial Parieto-Occipital Cortex of the Macaque Monkey," *Journal of Neuroscience* 30 (2010): 342–349.

55. Wolpert and Flanagan, "Motor Prediction."

Chapter 5

1. A. Straub, "The Effect of Lexical Predictability on Eye Movements in Reading: Critical Review and Theoretical Interpretation," *Language and Linguistic Compass* 9, no. 8 (2015): 311–327.

2. K. Rayner, T. J. Slattery, D. Drieghe, and S. P. Liversedge, "Eye Movements and Word Skipping during Reading: Effects of Word Length and Predictability," *Journal of Experimental Psychology: Human Perception and Performance* 37 (2011): 514–528.

3. Rayner et al., "Eye Movements and Word Skipping during Reading."

4. G. T. M. Altmann and Y. Kamide, "Incremental Interpretation at Verbs: Restricting the Domain of Subsequent Reference," *Cognition* 73 (1999): 247–264. Fig. 5.1 is adapted from E. B. Goldstein, *Cognitive Psychology*, 5th ed. (Boston: Cengage, 2019).

5. Fig. 5.2 is adapted from from K. A. De Long, T. P. Urbach, and M. Kutas, "Probabilistic Word Pre-activation during Language Comprehension Inferred from Electrical Brain Activity," *Nature Neuroscience* 8 (2005): 1117–1121; K. D. Federmeier, "Thinking Ahead: The Role and Roots of Prediction in Language Comprehension," *Psychophysiology* 44 (2007): 491–505; A. D. Patel and E. Morgan, "Exploring Cognitive Relations between Prediction in Language and Music," *Cognitive Science* 41 (2016): 303–320.

6. A. E. Kim, L. D. Oines, and L. Sikos, "Prediction during Sentence Comprehension Is More Than a Sum of Lexical Associations: The Role of Event Knowledge," *Language, Cognition, and Neuroscience* 31 (2015): 597–601.

7. C. Clifton, M. J. Traxler, M. T. Mohamed, R. S. Williams, R. K. Morris, and K. Rayner, "The Use of Thematic Role Information in Parsing:

Syntactic Processing Autonomy Revisited," *Journal of Memory and Language* 49 (2003): 317–334.

8. G. R. Kuperberg and T. F. Jaeger, "What Do We Mean by Prediction in Language Comprehension?" *Language, Cognition and Neuroscience* 31 (2015): 32–59.

9. T. G. Bever, "The Cognitive Basis for Linguistic Structures," in *Cognition and the Development of Language*, ed. J. R. Hayes (New York: Wiley, 1970), 279–362.

10. S. Koelsch, P. Vuust, and K. Friston, "Predictive Processes and the Peculiar Case of Music," *Trends in Cognitive Sciences* 23 (2018): P63–P77.

11. P. Vuust and M. A. G. Witek, "Rhythmic Complexity and Predictive Coding: A Novel Approach to Modeling Rhythm and Meter Perception in Music," *Frontiers in Psychology* 5 (2014): article 1111.

12. Figure 5.3 is adapted from T. Fujioka, L. J. Trainor, E. W. Large, and B. Ross, "Internalized Timing of Isochronous Sounds Is Represented in Neuromagnetic Beta Oscillations," *Journal of Neuroscience* 32 (2012): 1791–1802; Merchant et al., "Finding the Beat."

13. H. Merchant, J. Grahn, L. Trainor, M. Rohmeier, and W. T. Fitch, "Finding the Beat: A Neural Perspective across Humans and Non-human Primates," *Philosophical Transactions of the Royal Society B* 370 (2015): 20140093.

14. P. Vuust, M. J. Dietz, M. Witek, and M. L. Kringelbach, "Now You Hear It: A Predictive Coding Model for Understanding Rhythmic Incongruity," *Annals of the New York Academy of Sciences* 1423 (2018), 19–29.

15. Fig. 5.4d is adapted from P. Vuust, L. Ostergaard, K. J. Pallesen, C. Bailey, and A. Roepstorff, "Predictive Coding of Music: Brain Responses to Rhythmic Incongruity," *Cortex* 45 (2009): 80–92.

16. P. Janata, S. T. Tomic, and J. M. Haberman, "Sensorimotor Coupling in Music and the Psychology of the Groove," *Journal of Experimental Psychology: General* 14 (2011): 54–75; D. J. Levitin, J. A. Grahn, and J. London, "The Psychology of Music: Rhythm and Movement," *Annual Review of Psychology* 69 (2018).

17. Vuust et al., "Predictive Coding of Music."

18. S. Koelsch, P. Vuust, and K. Friston, "Predictive Processes and the Peculiar Case of Music," *Trends in Cognitive Sciences* 23 (2018): P63–P77; M. A. Rohrmeier and S. Koelsch, "Predictive Information Processing in Music Cognition: A Critical Review," *Internal Journal of Psychophysiology* 83 (2012): 164–175.

19. D. Deutsch, "Speaking in Tones," *Scientific American Mind*, July–August 2010, 36–43.

20. A. D. Patel, "Sharing and Nonsharing of Brain Resources for Language and Music," in *Language, Music, and the Brain*, ed. M. A. Arbib (Cambridge, MA: MIT Press, 2013), 329–355.

21. C. L. Krumhansl, "Perceiving Tonal Structure in Music," *American Scientist* 73 (1985): 371–378.

22. Fig. 5.5 is adapted from A. D. Patel, E. Gibson, J. Ratner, M. Besson, and P. J. Holcomb, "Processing Syntactic Relations in Language and Music: An Event-Related Potential Study," *Journal of Cognitive Neuroscience* 10 (1998): 717–733. See also S. Koelsch, S. Kilches, N. Steinbeis, and S. Schelinki, "Effects of Unexpected Chords and of Performer's Expression on Brain Responses and Electrodermal Activity," *PLoS One* 3, no. 7 (2007): e2631.

23. S. Koelsch, "Neural Substrates of Processing Syntax and Semantics in Music," *Current Opinion in Neurobiology* 15 (2005): 207–212; S. Koelsch, T. Gunter, A. D. Friederici, and E. Schroger, "Brain Indices of Music Processing: 'Nonmusicians' Are Musical," *Journal of Cognitive Neuroscience* 12 (2000): 520–541; B. Maess, S. Koelsch, T. C. Gunter, and A. D. Friederici, "Musical Syntax Is Processed in Broca's Area: An MEG Study," *Nature Neuroscience* 4 (2001): 540–545; Vuust et al., "Predictive Coding of Music."

24. A. R. Fogel, J. C. Rosenberg, F. M. Lehman, G. R. Kuperberg, and A. D. Patel, "Studying Musical and Linguistic Prediction in Comparable Ways: The Melodic Cloze Probability Method," *Frontiers in Psychology* 6 (2015): article 1718; data from fig. 7.

25. D. Gilbert and T. D. Wilson, "Why the Brain Talks to Itself: Sources of Error in Emotional Prediction," *Philosophical Transactions of the Royal Society B* 364 (2009): 1335–1341.

26. D. J. Simons and C. F. Chabris, "What People Believe about How Memory Works: A Representative Survey of the U.S. Population," *PLoS One* 6, no. 8 (2011): e22757.

27. F. C. Bartlett, *Remembering: A Study in Experimental and Social Psychology* (Cambridge: Cambridge University Press, 1932).

28. W. F. Brewer and J. C. Treyens, "Role of Schemata in Memory for Places," *Cognitive Psychology* 13 (1981): 207–230.

29. G. Kim, J. A. Lewis-Peacock, K. A. Norman, and N. B. Turk-Browne, "Pruning of Memories by Context-Based Prediction," *Proceedings of the National Academy of Sciences* 111 (2014): 8997–9002.

30. M. Vlascenau, R. Drach, and A. Coman, "Suppressing My Memories by Listening to Yours: The Triggered Context-Based Prediction Error on Memory," *Psychonomic Bulletin and Review* (2018), https://doi.org/10.3758/S13423-018-1481-2.

31. Gilbert and Wilson, "Why the Brain Talks to Itself."

32. D. L. Schacter and D. R. Addis, "The Cognitive Neuroscience of Constructive Memory: Remembering the Past and Imagining the Future," *Philosophical Transactions of the Royal Society of London B* 362 (2007): 773–786.

33. D. R. Addis, L. Pan, M.-A. Vu, N. Laiser, and D. L. Schacter, "Constructive Episodic Simulation of the Future and the Past: Distinct Subsystems of a Core Brain Network Mediate Imagining and Remembering," *Neuropsychologia* 47 (2009): 2222–2238; D. R. Addis, A. T. Wong, and D. L. Schacter, "Remembering the Past and Imagining the Future: Common and Distinct Neural Substrates during Event Construction and Elaboration," *Neuropsychologia* 45 (2007): 1363–1377.

34. Addis, Wong, and Schacter, "Remembering the Past"; D. Hassabis, D. Kumaran, S. D. Vann, and E. A. Maguire, "Patients with Hippocampal

Amnesia Cannot Imagine New Experiences," *Proceedings of the National Academy of Sciences* 104 (2007): 1726–1731; S. L. Mullally and F. A. Maguire, "Memory, Imagination and Predicting the Future: A Common Brain Mechanism?" *Neuroscientist* 20 (2014): 220–234.

35. D. L. Schacter, "Adaptive Constructive Processes and the Future of Memory," *American Psychologist* 67 (2012): 603–613.

36. E. C. Brown and M. Brüne, "The Role of Prediction in Social Neuroscience," *Frontiers in Human Neuroscience* 6 (2012): article 147, p. 12.

37. Brown and Brüne, "Role of Prediction," 12.

38. D. I. Tamir and M. A. Thornton, "Modeling the Predictive Social Mind," *Trends in Cognitive Sciences* 22 (2017): 201–212.

39. D. Premack and G. Woodruff, "Does the Chimpanzee Have a Theory of Mind?" *Behavior and Brain Sciences* 4 (1978): 515.

40. S. Baron-Cohen, A. M. Leslie, and U. Frith, "Does the Autistic Child Have a 'Theory of Mind'?" *Cognition* 21 (1985): 37–46.

41. M. Dolan and R. Fullam, "Theory of Mind and Mentalizing Ability in Antisocial Personality Disorders with and without Psychopathy," *Psychological Medicine* 34 (2004): 1093–1102; C. D. Frith and U. Frith, "How We Predict What Other People Are Going to Do," *Brain Research* 1079 (2006): 36–46; C. I. Hooker, S. C. Verosky, L. T. Germine, R. T. Knight, and M. D'Esposito, "Mentalizing about Emotion and Its Relationship to Empathy," *SCAN* 3 (2008): 204–217. The following draws a slight distinction between mentalizing and theory of mind: R. M. Carter and S. A. Hiettel, "A Nexus Model of the Temporal-Parietal Junction," *Trends in Cognitive Neuroscience* 17 (2013): 328–336.

42. R. Saxe and N. Kanwisher, "People Thinking about Thinking People: The Role of the Temporo-Parietal Junction in 'Theory of Mind,'" *NeuroImage* 19 (2003): 1835–1842.

43. Fig. 5.6 is adapted from fig. 2 of F. V. Van Overwalle and K. Baetens, "Understanding Others' Actions and Goals by Mirror and Mentalizing Systems: A Meta-analysis," *NeuroImage* 48 (2009): 564–584.

44. C. D. Frith and U. Frith, "The Neural Basis of Mentalizing," *Neuron* 50 (2006): 531–534; K. M. Ingelstrom, T. W. Webb, Y. T. Kelly, and M. S. A. Graziano, "Topographical Organization of Attentional, Social, and Memory Processes in the Human Temporoparietal Cortex," *eNeuro* 3, no. 2 (2016): e0060-16.2016 1–12; Koster-Hale and Saxe, "Theory of Mind"; R. Saxe, "Uniquely Human Social Cognition," *Current Opinion in Neurobiology* 16 (2006): 235–239; F. V. van Overwalle, "A Dissociation between Social Mentalizing and General Reasoning," *NeuroImage* 54 (2011): 1589–1599; F. V. Van Overwalle and K. Baetens, "Understanding Others' Actions and Goals by Mirror and Mentalizing Systems: A Meta-Analysis," *NeuroImage* 48 (2009): 564–584.

45. R. Kanai, B. Bahrami, R. Roylance, and G. Rees, "Online Social Network Size Is Reflected in Human Brain Structure," *Proceedings of the Royal Society B* 279 (2012): 1327–1334; J. Powell, P. A. Lewis, N. Roberts, M. Garcia-Fiñana, and R. I. M. Dunbar, "Orbital Prefrontal Cortex Volume Predicts Social Network Size: An Imaging Study of Individual Differences in Humans," *Proceedings of the Royal Society B* 279 (2012): 2157–2162.

46. F. Heider and M. Simmel, "An Experimental Study of Apparent Behavior," *American Journal of Psychology* 57 (1944): 243–259.

47. Castelli, F. Happe, U. Frith, and C. Frith, "Movement and Mind: A Functional Imaging Study of Perception and Interpretation of Complex Intentional Movement Patterns," *NeuroImage* 12 (2000): 314–325; A. Martin and J. Weisberg, "Neural Foundations for Understanding Social and Mechanical Concepts," *Cognitive Neuropsychology* 20 (2003): 575–587.

48. A. Cavallo, A. Koul, C. Ansuini, F. Capozzi, and C. Becchio, "Decoding Intentions from Movement Kinematics," *Science Reports* 6 (2016): 37036; A. Koul, A. Cavallo, F. Cauda, T. Costa, M. Diano, M. Pontil, and C. Becchio, "Action Observation Areas Represent Intentions from Subtle Kinematic Features," *Cerebral Cortex* 28 (2018): 2647–2654.

49. M. Iacoboni, I. Molnar-Szakacs, V. Gallese, G. Buccino, J. C. Mazziotta, and G. Rizzolatti, "Grasping the Intentions of Others with One's Own Mirror Neuron System," *PLoS Biology*, 3, no. 3 (2005): e79.

50. G. Di Pellegrino, L. Fadiga, L. Fogassi, V. Gallese, and G. Rizzolatti, "Understanding Motor Events: A Neurophysiological Study," *Experimental Brain Research* 91 (1992): 176–180; G. Rizzolatti and C. Sinigaglia, "The Mirror Mechanism: A Basic Principle of Brain Function," *Nature Reviews Neuroscience*, 17 (2016): 757–765.

51. Frith and Frith, "How We Predict What Other People Are Going to Do"; C. Keysers, B. Wicker, V. Gazzola, J.-L. Anton, L. Fogassi, and V. Gallese, "A Touching Sight: SII/PV Activation during the Observation and Experience of Touch," *Neuron* 42 (2004): 335–346; T. Singer, B. Seymour, J. O'Doherty, H. Kaube, R. J. Dolan, and C. D. Frith, "Empathy for Pain Involves the Affective but Not Sensory Components of Pain," *Science* 303 (2004): 1157–1162.

52. Koul et al., "Action Observation Areas Represent Intentions."

53. Iacoboni et al., "Grasping the Intentions of Others."

54. V. Gallese and C. Sinigaglia, "What Is So Special about Embodied Simulation?" *Trends in Cognitive Sciences*, 15 (2011): 512–519; G. Rizzolatti and C. Sinigaglia, "The Mirror Mechanism: A Basic Principle of Brain Function," *Nature Reviews Neuroscience* 17 (2016): 757–765; R. Saxe, "Uniquely Human Social Cognition,." *Current Opinion in Neurobiology* 16 (2006): 235–239.

55. L. Cattaneo and G. Rizzolatti, "The Mirror Neuron System," *Neurological Review* 66 (2009): 557–560; Rizzolatti and Sinigaglia, "The Mirror Mechanism." Describing doubts about mirror neurons: G. Hickock, (2008). "Eight Problems for the Mirror Neuron Theory of Action Understanding in Monkeys and Humans," *Journal of Cognitive Neuroscience* 21 (2008): 1229–1243; B. Thomas, "What's So Special about Mirror Neurons?" *Scientific American*, November 6, 2012

56. R. Lemon, "Is the Mirror Cracked?" *Brain* 138 (2015): 2109–2110.

57. R. P. Spunt and M. D. Lieberman, "The Busy Social Brain: Evidence for Automaticity and Control in the Neural Systems Supporting Social Cognition and Action Understanding," *Psychological Science* 24 (2013): 80–86.

58. Koster-Hale and Saxe, "Theory of Mind."

59. A. Clark, "A Nice Surprise? Predictive Processing and the Active Pursuit of Novelty," *Phenomenology and Cognitive Science* 17 (2018): 521–534; J. Kiverstein, M. Miller, and E. Rietveld, "The Feeling of Group: Novelty, Error Dynamics, and the Predictive Brain," *Synthese* (2017): 1–23.

60. S. Koelsch, P. Vuust, and K. Friston, "Predictive Processes and the Peculiar Case of Music," *Trends in Cognitive Sciences* 23 (2019): 6377.

Chapter 6

1. C. S. Sherrington, *Man and His Nature* (New York: Cambridge University Press, 1942), 178.

2. P. Corsi, *The Enchanted Loom: Chapters in the History of Neuroscience* (New York: Oxford University Press, 1991); R. Cotterill, *Enchanted Looms: Conscious Networks in Brains and Computers* (New York: Cambridge University Press, 1998); R. Jastrow, *The Enchanted Loom: Mind in the Universe* (New York: Simon & Schuster, 1981). The origin of the phrase "enchanted loom" is interesting, because it is likely that it and the phrase "flashing shuttles" were inspired by a nineteenth-century weaving device called the Jacquard loom, which was designed to weave complex patterns and was controlled by a system of punch cards that were the forerunners of the punch cards used to program computers until the 1970s. S. Finger, *Minds behind the Brain* (New York: Oxford University Press, 1999).

3. O. Sporns, "Cerebral Cartography and Connectomics," *Philosophical Transactions of the Royal Society* B370 (2015): 1–12, 2.

4. O. Sporns, G. Tononi, and R. Kotter, "The Human Connectome: A Structural Description of the Human Brain," *PLoS Computational Biology* 1, no. 4 (2005): 245–251.

5. Sporns, Tononi, and Kotter, "The Human Connectome," 245.

6. A. Baronchelli, R. Ferrer-i-Cancho, R. Pastor-Satorras, N. Chater, and M. H. Christiansen, "Networks in Cognitive Science," *Trends in Cognitive Sciences* 17, no. 7 (2013): 348–360, 349.

7. M. F. Glaser, S. M. Smith, D. S. Marcus, J. L. R. Anderson, E. J. Aurbach, T. E. J. Behrens, T. S. Coalson, M. P. Harms., M. Jenkinson, S. Moeller, et al. "The Human Connectome Project's Neuroimaging Approach," *Nature Neuroscience* 19, no. 9 (2016): 1175–1187.

8. Fig. 6.1 is from F. Calamante, R. A. J. Masterton, J. D. Tournier, R. E. Smith, L. Willats, D. Raffelt, and A. Connelly, "Track-Weighted Functional Connectivity (TW-FC): A Tool for Characterizing the Structural-Functional Connections in the Brain," *NeuroImage* 70 (2013): 199–210. See also S. L. Bressler and V. Menon, "Large-Scale Brain Networks in Cognition: Emerging Methods and Principles," *Trends in Cognitive Sciences* 14 (2010): 277–290; O. Sporns, "Cerebral Cartography and Connectomics," *Philosophical Transactions of the Royal Society B* 370 (2015): 20140173.

9. Fig. 6.2 is adapted from fig. 2.27 in Goldstein, *Cognitive Psychology*, 5th ed. See also B. Biswal, F. Z. Yetkin, V. M. Haughton, and J. S. Hyde, "Functional Connectivity in the Motor Cortex of Resting Human Brain Using Echo-Planar MRI," *Magnetic Resonance in Medicine* 34 (1995): 537–541.

10. Fig. 6.3 is adapted from D. L. Zabelina and J. R. Andrews-Hanna, "Dynamic Network Interactions Supporting Internally-Oriented Cognition," *Current Opinion in Neurobiology* 40 (2016): 96–93. Table 6.1 is adapted from Zabelina and Andrews-Hanna, "Dynamic Network Interactions Supporting Internally-Oriented Cognition"; D. M. Barch, "Brain Network Interactions in Health and Disease," *Trends in Cognitive Sciences* 17, no. 12 (2013): 603–605; Bressler and Menon, "Large-Scale Brain Networks in Cognition"; R. L. Buckner and F. M. Krienen, "The Evolution of Distributed Association Networks in the Human Brain," *Trends in Cognitive Sciences* 17 (2013): 648–665; M. E. Raichle, "The Restless Brain," *Brain Connectivity* 1, no. 1 (2011): 3–12.

11. G. Zeng, D. Li, S. Guo, L. Gao, Z. Gao, H. E. Stanley, and S. Havlin, "Switch between Critical Percolation Modes in City Traffic Dynamics," *Proceedings of the National Academy of Sciences* 116, no. 1 (2019): 23–28.

12. M. P. Van den Heuvel and H. E. H. Pol, "Exploring the Brain Network: A Review on Resting-State fMRI Functional Connectivity," *European Neuropsychopharmacology* 20 (2010): 519–534.

13. D. S. Bassett and O. Sporns, "Network Neuroscience," *Nature Neuroscience* 20 (2017): 353–364; O. Sporns, "Graph Theory Methods: Applications in Brain Networks," *Dialogues in Clinical Neuroscience* 20 (2018): 111–120; G. S. Wig, B. L. Schlaggar, and S. E. Petersen, "Concepts and Principles in the Analysis of Brain Networks," *Annals of the New York Academy of Sciences* 1224 (2011): 126–146.

14. Figs. 6.4b and 6.4c are based on fig. 10 in K. J. Friston, "Functional and Effective Connectivity: A Review," *Brain Connectivity* 1 (2011): 32.

15. B. J. Shannon, A. Desenbach, Y. Su., A. G. Vlassenko, L. J. Larson-Prior, T. S. Nolan, A. Z. Snyder, and M. E. Raichle, "Morning-Evening Variation in Human Brain Metabolism and Memory Circuits," *Journal of Neurophysiology* 109 (2013): 1444–1456.

16. R. A. Poldrack, T. O. Laumann, O. Koyejo, B. Gregory, A. Hover, M.-Y. Chen, K. J. Gorgolewski, J. Luci, S. J. Joo, R. L. Boyd, et al., "Long-Term Neural and Physiological Phenotyping of a Single Brain," *Nature Communications* 6 (2015): 1–15. Fig. 6.5 is adapted from fig. 4 in this paper.

17. M. E. Raichle, A. M. MacLeod, A. Z. Snyder,W. J. Powers, D. A. Gusnard., and G. L. Shulman, "A Default Mode of Brain Function," *Proceedings of the National Academy of Sciences* 98, no. 2 (2001): 676–682.

18. M. A. Killingsworth and D. T. Gilbert, "A Wandering Mind Is an Unhappy Mind," *Science* 330 (2010): 932.

19. E. Barron, L. M. Riby, L. Greer, and J. Smallwood, "Absorbed in Thought: The Effect of Mind Wandering on the Processing of Relevant and Irrelevant Events," *Psychological Science* 22, no. 5 (2011): 596–601; C. Gil-Jardiné, M. Née, E. Lagarde, J. Schooler, B. Contrand, L, Orriols, and C. Galera, "The Distracted Mind on the Wheel: Overall Propensity to Mind Wandering Is Associated with Road Crash Responsibility," *PLOS One* 12, no. 8 (2017): 1–10; J. Smallwood, "Mind-Wandering While Reading: Attentional Coupling, Mindless Reading and the Cascade Model of Inattention," *Language and Linguistics Compass* 5 (2011): 63–77; J. Smallwood, E. Beach, J. W. Schooler, and T. Handy, "Going AWOL in the Brain: Mind Wandering Reduces Cortical Analysis of External Events," *Journal of Cognitive Neuroscience* 20, no. 3 (2008): 458–469.

20. B. Baird, J. Smallwood, M. D. Mrazek, J. W. Y. Kam, M. S. Franklin, and J. W. Schooler, "Inspired by Distraction: Mind Wandering Facilitates Creative Incubation," *Psychological Science* 23, no. 10 (2012): 1117–1122.

21. M. W. Franklin, M. D. Mrazek, C. L. Anderson, J. Smallwood, A. Kingstone, and J. W. Schooler, "The Silver Lining of a Mind in the Clouds: Interesting Musings Are Associated with Positive Mood While Mind-Wandering," *Frontiers in Psychology* 4 (2013): article 583; G. L. Poerio, P. Totterdell, L.-M. Emerson, and E. Miles, "Love Is the Triumph of the Imagination: Daydreams about Significant Others Are Associated with Increased Happiness, Love and Connection," *Consciousness and Cognition* 33 (2015): 135–144.

22. B. Baird, J. Smallwood., and J. W. Schooler, "Back to the Future: Autobiographical Planning and the Functionality of Mind-Wandering," *Consciousness and Cognition* 20 (2011): 1604–1611; D. Stawarczyk, H. Cassol, and A. D'Argembeau, "Phenomenology of Future-Oriented Mind-Wandering Episodes," *Frontiers in Psychology* 4 (2013): article 425.

23. K. Christoff, A. M. Gordon, J. Smallwood, R. Smith, and J. Schooler, "Experience Sampling during fMRI Reveals Default Network and Executive System Contributions to Mind Wandering," *Proceedings of the National Academy of Sciences* 106 (2009): 8719–8724.

24. R. E. Beaty, M. Benedek, R. W. Wilkins, E. Jauk, A. Fink, P. J,. Silviam, D. A. Hodges, K. Koschutnig, and A. C. Neubauer, "Creativity and the Default Network: A Functional Connectivity Analysis of the Creative Brain at Rest," *Neuropsychologia* 64 (2014): 94.

25. N. Bar, "Intuition, Reason and Creativity," in *The New Reflectionism in Cognitive Psychology: Why Reason Matters*, ed. G. Pennycook (Routledge, 2018); R. E. Jung, B. S. Mead, J. Carrasco, and R. A. Flores, "The Structure of Creative Cognition in the Human Brain," *Frontiers in Human Neuroscience* 7 (2013): article 330; N. Mayseless, A. Eran, and S. G. Shamay-Tsoory, "Generating Original Ideas: The Neural Underpinnings of Originality," *NeuroImage* 116 (2015): 232–239.

26. Beaty et al., "Creativity and the Default Network," 94; R. E. Beaty, M. Benedek, S. B. Kaufman, and P. J. Silvia, "Default and Executive

Network Coupling Supports Creative Idea Production," *Scientific Reports* 5 (2015): 1–14.

27. Beaty et al., "Creativity and the Default Network," 94.

28. R. E. Beaty, M. Benedek, P. J. Silvia, and D. L. Schacter, "Creative Cognition and Brain Network Dynamics," *Trends in Cognitive Sciences* 20, no. 2 (2016): 87–95.

29. M. Ellamil, C. Dobson, M. Beeman, and K. Christoff, "Evaluative and Generative Modes of Thought during the Creative Process," *NeuroImage* 59 (2012): 1783–1794.

30. Zabelina and Andrews-Hanna, "Dynamic Network Interactions"; Beaty et al., "Creative Cognition and Brain Network Dynamics"; Ellamil et al., "Evaluative and Generative Modes of Thought."

31. Beaty et al., "Default and Executive Network Coupling." Fig. 6.6 is adapted from fig. 6 in this paper.

32. R. Schmälzle, M. B. O'Donnell, J. O. Garia, C. N. Cascio, J. Bayer, D. S. Bassett, J. M. Vettel, and E. B. Falk, "Brain Connectivity Dynamics during Social Interaction Reflect Social Network Structure," *Proceedings of the National Academy of Sciences* 114, no. 20 (2017): 5153–5158.

33. Fig. 6.7 is adapted from fig. 1 (top) in Schmälzle et al., "Brain Connectivity Dynamics."

34. N. I. Eisenberger, "The Pain of Social Disconnection: Examining the Shared Neural Underpinnings of Physical and Social Pain," *Nature Reviews Neuroscience* 13 (2012): 421–434.

35. D. Alcala-Lopez, J. Smallwood, E. Jefferies, F. Van Overwalle, K. Vogeley, R. B. Mars, B. I. Turetsky, et al., "Computing the Social Brain Connectome across Systems and States," *Cerebral Cortex* 28 (2018): 2207–2232; Kanai et al., "Online Social Network Size."

36. J. Powell, P. A. Lewis, N. Roberts, M. Garcia-Finana, and R. I. M. Dunbar, "Orbital Prefrontal Cortex Volume Predicts Social Network Size: An Imaging Study of Individual Differences in Humans," *Proceedings of the Royal Society B* 279 (2012): 2157–2162.

37. U. Hassan, A. A. Ghazanfar, B. Galantucci, S. Garrod, and C. Keysers, "Brain-to-Brain Coupling: A Mechanism for Creating and Sharing a Social World," *Trends in Cognitive Sciences* 16 (2012): 114–121; M. D. Liberman, "Birds of a Feather Synchronize Together," *Trends in Cognitive Sciences* 22 (2018): 371–372; C. Parkinson, A. M. Kleinbaum, and T. Wheatley, "Similar Neural Responses Predict Friendship," *Nature Communications* 9, no. 322 (2018): 1–14.

38. L. Geerligs, R. J. Benken, E. Saliasi, N. M. Maurits, and M. M. Lorist, "A Brain-Wide Study of Age-Related Changes in Functional Connectivity," *Cerebral Cortex* 25 (2015): 1987–1999.

39. Beaty et al., "Creativity and the Default Network," 94; Beaty et al., "Default and Executive Network Coupling."

40. Schmälzle et al., "Brain Connectivity Dynamics."

41. Shannon et al., "Morning-Evening Variation."

42. Poldrack et al., "Long-Term Neural and Physiological Phenotyping."

43. S. Y. Bookheimer, D. H. Salat, M. Terpstra, B. M. Ances, D. M, Barch, R. L. Buckner, G. C. Burgess, et al., "The Lifespan Human Connectome Project in Aging: An Overview," *NeuroImage* 185 (2019): 335–348.

44. R. F. Betzel, L. Byrge, Y. H. Joaquin Goni, X.-N. Zuo, and O. Sporn, "Changes in Structural and Functional Connectivity among Resting-State Networks across the Human Lifespan," *NeuroImage* 102 (2014): 345–357.

45. C. J. Honey, R. Kotter, M. Breakspear, and O. Sporns, "Network Structure of Cerebral Cortex Shapes Functional Connectivity on Multiple Time Scales," *Proceedings of the National Academy of Sciences* 104, no. 24 (2007): 10240–10245; C. J. Honey, O. Sporns, L. Cammoun, X. Gigandet, J. P. Thiran, R. Meuli, and P. Hagmann, "Predicting Human Resting-State Functional Connectivity from Structural Connectivity," *Proceedings of the National Academy of Sciences* 106, no. 6 (2009): 2035–2040.

46. M. Y. Chan, D. C. Park, N. K. Savilia, S. E. Petersen, and G. S. Wig, "Decreased Segregation of Brain Systems across the Healthy Adult Life

Span," *Proceedings of the National Academy of Sciences* 111, no. 46 (2014): E4977–E5006.

47. Betzel et al., "Changes in Structural and Functional Connectivity"; J. S. Damoiseaux, "Effects of Aging on Functional and Structural Brain Connectivity," *NeuroImage* 160 (2017): 32–40; L. Z. Ferreira, A. C. B. Regina, N. Kovacevic, M. daG, M. Martin., P. P. Santos, G. de G. Carneiro, et al., "Aging Effects on Whole-Brain Functional Connectivity in Adults Free of Cognitive and Psychiatric Disorders," *Cerebral Cortex* 26 (2016): 3851–3865; D. C. Park and P. Reuter-Lorenz, "The Adaptive Brain: Aging and Neurocognitive Scaffolding," *Annual Review of Psychology* 60 (2009): 173–196.

48. Fig. 6.8 is adapted from Chan et al., "Decreased Segregation of Brain Systems across the Healthy Adult Life Span."

49. G. R. Turner and R. N. Spreng, "Prefrontal Engagement and Reduced Default Network Suppression Co-occur and Are Dynamically Coupled in Older Adults: The Default-Executive Coupling Hypothesis of Aging," *Journal of Cognitive Neuroscience* 27, no. 12 (2015): 2462–2476.

50. Bookheimer et al., "Lifespan Human Connectome Project."

51. F.-C. Yeh, J. M. Vettel, A. Singh, B. Poczos, S. T. Grafton, K. I. Erickson, W-Y. I. Tseng, and T. D. Verstynen, "Quantifying Differences and Similarities in Whole-Brain White Matter Architecture Using Local Connectome Fingerprints," *PLOS Computational Biology* 12, no. 11 (2016): e1005203.

52. Geshe Kelsang Gyatso, *Transform Your Life* (Ulverston, UK: Tharpa, 2002).

Further Reading

General

Brockman, J. *The Mind*. HarperCollins, 2011. Essays on various aspects of the mind.

Eagleman, D. *The Brain: The Story of You*. Pantheon, 2015. Popular treatment based on the PBS series.

Gregory, R. *Eye and Brain*. 5th ed. Princeton University Press, 1997. Classic popular book focusing on visual perception.

Gregory, R. *Oxford Companion to the Mind*. 2nd ed. Oxford University Press, 2004. An encyclopedic volume of over one thousand pages with alphabetical entries by over three hundred authors.

Gregory, R. *Seeing through Illusions*. Oxford, 2009. What illusions tell us about how the brain perceives the world.

Pinker, S. *How the Mind Works*. Norton, 1997. Overview of the mind's operation.

Ramachandran, V. S. *The Tell-Tale Brain: A Neuroscientist's Quest for What Makes Us Human*. Norton, 2011. A book focusing largely on neuropsychological case studies.

Ramachandran, V. S., and S. Blakeslee. *Phantoms in the Brain: Probing the Mysteries of the Human Mind*. William Morrow, 1998. Interesting accounts of neuropsychological case studies.

Sacks, O. *The Man Who Mistook His Wife for a Hat*. Picador, 1986. Fascinating cases of brain damage made accessible to the general reader.

Wenk, G. L. *The Brain: What Everyone Needs to Know*. Oxford University Press, 2017. A series of 112 questions relevant to what the brain does.

More Focused

Ariely, D. *Predictably Irrational: The Hidden Forces That Shape Our Decisions*. HarperCollins, 2009.

Barrett, L. F. *How Emotions Are Made: The Secret Life of the Brain*. Houghton Mifflin Harcourt, 2017.

Beilock, S. *How the Body Knows Its Mind: The Surprising Power of the Physical Environment to Influence How You Think and Feel*. Simon & Schuster, 2015.

Blackmore, S. *Conversations on Consciousness*. Oxford University Press, 2005. A number of experts chime in.

Davidson, R. J., and S. Begley. *The Emotional Life of Your Brain: How Its Unique Patterns Affect the Way You Think, Feel and Live—and How You Can Change Them*. Hudson Street Press, 2012.

Duhring, C. *The Power of Habit*. Random House, 2012. Focuses on habitual behaviors in personal life and business practices.

Editors of *Scientific American*. *Remember When: The Science of Memory*. Scientific American, 2013. A selection of interviews and articles about various aspects of memory.

Editors of *Scientific American*. *The Secrets of Consciousness*. Scientific American, 2013. Articles from *Scientific American* that focus on consciousness.

Engelman, D. *Incognito: The Secret Lives of the Brain*. Pantheon, 2011. Wide-ranging description of the experiences and behaviors associated with unconscious brain processes.

Evans, Jonathan St. B. T. *Thinking and Reasoning: A Very Short Introduction*. Oxford University Press, 2017.

Fernyhough, C. *Pieces of Light: How the New Science of Memory Illuminates the Stories We Tell about Our Pasts*. HarperCollins, 2013.

Gazzaniga, M. *Who's in Charge? Free Will and the Science of the Brain*. HarperCollins, 2011.

Gigerenzer, G. *Risk Savvy: How to Make Good Decisions*. Oxford University Press, 2014.

Goldberg, E. *The New Executive Brain: The Frontal Lobes in a Complex World*. Oxford University Press, 2009. Goldberg's view of the functioning of the frontal lobes with respect to executive functions.

Iacoboni, M. *Mirroring People: The New Science of How We Connect with Others*. Farrar, Straus and Giroux, 2009.

Jackendoff, R. *A User's Guide to Thought and Meaning*. Oxford University Press, 2012.

Johnston, E., and L. Olson. *The Feeling Brain: The Biology and Psychology of Emotions*. Norton, 2015.

Kahneman, Daniel. *Thinking Fast and Slow*. Farrar, Straus and Giroux, 2012. Focuses on problem-solving, reasoning, and decision-making.

Kandel, E. R. *In Search of Memory*. Norton, 2006.

Klingberg, T. *The Overflowing Brain*. Oxford University Press, 2009.

Lieberman, M. D. *Social: Why Our Brains Are Wired to Connect*. Crown, 2013.

Markman, Art. *Smart Thinking: Three Essential Keys to Solve Problems, Innovate, and Get Things Done*. Penguin, 2012. Focuses on learning, problem-solving, and creativity.

Mischell, W. *The Marshmallow Test*. Little, Brown, 2014. Relevant to self-control.

Montgomery, S. *The Soul of an Octopus: A Surprising Exploration into the Wonder of Consciousness*. Atria, 2015.

Nisbett, R. *Mindware: Tools for Smart Thinking.* Farrar, Straus and Giroux, 2015.

Pessoa, L. *The Cognitive-Emotional Brain.* MIT Press, 2013.

Redish, A. D. *The Mind within the Brain: How We Make Decisions and How Those Decisions Go Wrong.* Oxford University Press, 2013.

Robison, J. E. *Switched On: A Memoir of Brain Change and Emotional Awakening.* Spiegel & Grau, 2016.

Schacter, D. L. *The Seven Sins of Memory.* Houghton Mifflin, 2001.

Name Index

Addis, Rose, 139
Adrian, Edgar, 27, 29
Alexander, Eben, 22

Baird, Benjamin, 169
Bar, Moshe, 102, 105
Barrett, L. F., 102
Barry, Susan, 47–49
Bartlett, Fredric, 137
Bassett, Danielle, 58
Bateson, Melissa, 90
Bayes, Thomas, 102–104
Beaty, Roger, 170–172
Berger, Hans, 27, 29
Bever, Thomas, 128
Bird, Christopher, 35
Blakemore, Sarah-Jayne, 116
Blakeslee, Sandra, 70
Broadbent, Donald, 13–14
Broca, Paul, 24–25, 155
Brockman, John, 24
Brown, Elliot, 140
Brüne, Martin, 140

Chalmers, David, 32, 41, 44, 52–54
Chamovitz, Daniel, 36
Chan, Micaela, 176–178
Cherry, Colin, 13–14
Chomsky, Noam, 16
Chopra, Deepak, 20
Clark, Andy, 109
Corsi, Pietro, 157

Descartes, René, 19–20, 34
Donders, Franciscus, 7–9

Ebbinghaus, Hermann, 9
Einstein, Albert, 169
Eisenhower, Dwight, 17

Fried, Itzhak, 80

Gallant, Jack, 56
Gazzaniga, Michael, 58
Geerligs, Linda, 176
Gilbert, Daniel, 136, 139, 168
Gopnik, Alison, 56

Gregory, Richard, 98–99
Gyatso, Geshe Kelsang, 20, 181

Haggard, Patrick, 83
Hartline, Keffer, 97
Heider, Fritz, 145–146
Helmholtz, Hermann von, 94,
   98–99, 104, 112, 128
Hitchcock, Alfred, 1
Holland, Rob, 91
Hubel, David, 60, 97
Huxley, Thomas, 55

Iacoboni, Mario, 147–148

Jackson, Frank, 46
James, William, 33
Jastrow, Robert, 157

Kay, Aaron, 90
Killingsworth, Matthew, 168
King, Martin Luther, 65
Koch, Christof, 43
Koelsch, Stefan, 129
Kok, Peter, 94
Koster-Hale, Jorie, 151
Koul, Atesh, 146–148

Leary, Timothy, 18
Levine, Joshph, 55
Libet, Benjamin, 77–84

May, Michael, 102
McCarthy, John, 14
Meyer, Travis, 107
Miller, George, 15

Miller, Jeff, 81–83
Milner, David, 68–69
Molaison, Henry, 59
Mozart, Amadeus, 153–154

Naci, Lorina, 1
Nagel, Thomas, 41
Neisser, Ulrich, 17
Newell, Alan, 15
Newton, Isaac, 50, 169

Olson, Carl, 107

Patel, Aniruddh, 133–135
Pavlov, Ivan, 11
Poincaré, Henri, 169
Premack, David, 142

Raicle, Marcus, 168
Ramachandran, V. S., 70
Ramón y Cajal, Santiago,
   25–27, 29
Rayner, Keith, 123
Rayner, Rosalie, 10–11
Reber, Arthur, 85–86
Riddoch, George 71

Sacks, Oliver, 23, 47–48, 66–67
Saffran, Jennifer, 87–89
Saxe, Rebecca, 151
Schacter, Stanley, 139
Schmälzle, Ralf, 173–174
Schwarz, Wolf, 81–83
Seth, Anil, 32
Sherrington, Charles, 156
Simmel, Marianne, 145–146

Name Index

Simon, Herb, 15
Skinner, B. F., 11–12, 16
Smith, Adam, 17
Soon, Chun Siong, 80, 82–83
Sporns, Olaf, 157

Thompkins, Peter, 35
Tononi, Giulio, 43

Van Gulick, Robert, 31

Watson, John B., 10–11
Weiskrantz, Lawrence, 72
Wernicke, Carl, 25, 29, 155
Wiesel, Torsten, 60, 97
Wilson, Timothy, 136, 139
Woodruff, Greg, 142
Wundt, Wilhelm, 9

Yeh, Fang-Cheng, 178

Subject Index

Page numbers followed by an "f" or "t" indicate figures or tables, respectively.

Adaptive functions
  of language perception, 128
  of memory, 140
  of music prediction, 139, 152
  of other-touching perception, 115
  of prediction, 107, 152, 154
  of response to prediction error, 107
Afterlife hypothesis, 22
Alternate uses task, 169, 171
Ambiguity
  of language, 5
  of retinal image, 5, 94–97
Amnesia, and priming, 90
Amygdala, 59–60
Analytic introspection, 9
  and brain damage, 66–76
  and decision making, 77–84
  and implicit learning, 84–89
  and priming, 89–91

Animal experience, 40–42
Anthromorphism, 37, 190n19
Artificial intelligence 14–15
Attention, 14, 15
  and neglect, 73–76
  task-related, 111–112
Attributing intentions to geometrical objects, 145–146

Bayesian inference, 102–104
Bayes' theorem, 102–104
Behaviorism, 10–11
Binocular depth perception, experience of 47–49
Blindsight, 71–73
Bottom-up processing, 98–99, 104
Brain. *See also* Brain damage; Brain structures
  communication system, 157–170, 178–181
  connections to mind, 24–30

Brain (cont.)
  and conscious experience, 38–41, 51–63
  creation of the mind, 155
  and creative thinking, 170–173
  and decision making, 77–84
  and memory in aging, 176–178
  and prediction, 94, 105–119, 125–126, 129–130, 133–135, 139–140, 143–145, 147–151
  as a prediction machine, 94
  as a probability computer, 98
  skepticism about role in mind, 19–23
  and social cognition, 173–176
Brain damage, 66–76
  and blindsight, 71–73
  and language, 24–25
  and memory, 140
  occipital lobe damage, 71–72
  parietal lobe damage, 69–71
  temporal lobe damage, 69–71
Brain imaging. *See* Functional magnetic resonance imaging
Brain structures, 39f, 70f
  amygdala, 59–60
  Broca's area, 25–26, 58
  connectome, 157–159
  default network, 161–163, 168–172
  dorsal attention network, 161–163
  dorsal pathway, 68–71
  executive control network, 161–163, 170–172
  feature detectors, 97
  fusiform face area, 61–62
  hippocampus, 59, 176
  intraparietal sulcus and mirror network, 144–145
  medial temporal cortex, 55–56
  medial temporal lobe and memory, 166
  mentalizing network, 144
  mirror network, 144
  mirror neurons, 147–149
  occipital lobe damage, 71–72
  parietal lobe damage, 69–71
  parietal reach region, 118
  prefrontal cortex, 105
  prefrontal cortex and mentalizing network, 145, 175
  premotor cortex and mirror network, 144, 145
  prefrontal cortex and social networks, 175
  right-hemisphere damage, 73
  salience network, 161–162
  somato-motor network, 161–162
  somatosensory cortex and mirror network, 148
  superior colliculus, 73
  superior temporal sulcus and mirror network, 144–145
  temporal cortex, 105
  temporal lobe damage, 69–71
  temporal parietal junction (TBJ) and mentalizing, 144–145
  visual network, 161–162
  Wernicke's area, 25–26, 58

Subject Index

what pathway, 68–71
where/how pathway, 68–71
Broca's area, 25–26, 58

Cartesian dualism, 19
Cloze probability
   and language, 123–125
   and music, 135
Cognitive psychology textbook, 17
Cognitive revolution, 16–17
Color experience, 50–51
Computer flow diagram, 13
Conditioning, 11–12
Connectivity, 155–178. *See also* Functional connectivity
   and aging, 176–178
   and creative thinking, 170–173
   and fasting, 166–167
   functional, 159–178
   and functional networks, 159–165
   and network interactions, 167–170
   and social cognition, 173–176
   structural, 157–158
Connectome, 157–159
Consciousness, 1–61
   in bats, 41, 42
   in coma, 1–3
   definitions, 31–33
   easy problem, 56, 155
   and experience, 31–63
   first-person approach, 32
   hard problem, 52–57, 155
   in nonhuman animals, 34–35, 37–42
   in plants, 35–36
Constructive episodic simulation hypothesis, 139
Cooperation, unconscious influences, 90–91
Corollary discharge, 112–119
   and the comparator, 204n46
   and cricket perception, 114
   and eye movements, 113–114
   and grasping and gripping, 119
   and touch and tickle, 115–116
Creation of experience. *See* Nervous system

Decision making, 77–84
   and mind, 4, 7
Dedifferentiation of modules, 177–179
Default mode network, 161–163, 168–170
   interaction with executive control network, 171–172
Determinism, 83
Distributed processing, 155
Distributed representation, 61–63
Dorsal pathway, 68–71
Dynamic highways of the mind, 155–181
Dynamics of cognition, 165–177

Efference copy, 112–113. *See* Corollary discharge
Electroencephalogram, 27, 29, 77–79

Emotional reactions, 59–60
Enchanted Loom (Sherrington), 156–157, 212n2
Event-related potential, 125–126, 135
Executive control network, 170–172
Experience. *See also* Consciousness; Qualia
 and analytic introspection, 9–10
 and brain damage, 46–77
 of color, 50–51
 in coma, 1–3
 and consciousness, 31–63
 creation by nervous system, 24–30, 49–63
 of depth, 46–49
 and drugs, 18
 and eye movements, 110–114
 fairy, 43
 near-death, 22–23
 out-of-body, 21
 as private, 32–33, 43–46
 and touching, 115–116
Explanatory gap, 55
Extinction, 74–76
Eye and Brain (Gregory), 98–99
Eye movements, 110–114

False belief task, 143–144
Feature detectors, 97
Fixations, 111
Free will, 80–81, 83
Functional connectivity, 159–161, 162t, 163–179, 180t, 181
 and aging, 176–178
 and creative thinking, 170–173
 and fasting, 166–167
 measurement of, 159–160
 and memory, 166
 networks, 157–162
 and social inclusion, 173–175
 temporal connectivity analysis, 172
Functional magnetic resonance imaging (fMRI), 27–30
 in coma patient, 1–2
 and memory, 139
 and neural mind reading, 56–57
 resting state, 159–162
Functional organization, 58–63
Functions of mind, 4–5
Fusiform face area, 61–62

Garden path sentence, 127
Grammar
 and music, 135
 unconscious use of, 84–89
Graph theory, 163–165

Hallucinations, 23
Hard problem of consciousness, 52–56
Hidden mechanisms of mind, 2–3, 6, 65–92
Hierarchical predictive processing, 108–109
Hierarchical processing, 97
Hippocampus, 59, 176
Hippocampus, decrease with aging, 176

Subject Index

Human experience, private nature of, 42–49
Human potential movement, 19
Hypotheses and object perception, 98

Implicit learning, 84–89
Incubation, and mind wandering, 169
Infant learning, 87–88
Inference, 8–9, 94–97
  from actions, 145–147
  Bayesian, 102–104
  Donders, 8–9
  and prediction, 137
  and social cognition, 173–176
  unconscious (Helmholtz), 94–104, 109
  theory of mind, 142–147
Information flow within networks, 161–163
Information processing approach to the mind, 13–14
Intelligent behavior in animals, 37–38
Interacting with the environment, 109–119
Intraparietal sulcus and mirror network, 144–145
Inverse projection problem, 96

Knowledge
  and eye movements, 112
  and hidden mechanisms, 6
  in machines, 99
  and memory, 136–138
  and qualia, 46–49
  and scene perception, 65–66
  and top-down processing, 98–99, 104
  unconscious, 70, 84–92

Language, 85–89
Language development, 16
Libet's experiment, 77–80
Light from above assumption, 99, 101
Likelihood principle, 96, 128
Localization of function, 58–61, 155, 159
LSD, 18, 23

Mary the color scientist, 46–48
Measurement of mind, 7–8
Medial temporal cortex, 55–56
Medial temporal lobe and memory, 166
Meditation, 18
Memory
  construction, 136–138
  decrease with aging, 178
  and hippocampus, 62–63
  and mind, 4
  predictive editing, 138–139
  schemas, 137
  simulating the future, 139–140
Mental health and mind, 4
Mentalizing, 142–147
  dual-process theory, 149
Mentalizing network, 144
  and social cognition, 173–175
Mind, definition, 4

Mind-body separation, 20, 21
Mind-brain connections, 24–30, 63
Mind-brain skepticism, 19–24
Mind flow diagram, 13, 14
Mind wandering, 168–170
Mirror network, 144
Mirror neurons, 147–149
and pain, 148
Modularity, decrease with aging, 177–179
Motion perception, 55–56
Movement and prediction, 109–119
Multifaceted nature of mind, 4
Music, predicting timing, 129–131
Musical syntax, 131, 133

Near-death experience, 22–23
Nervous system, creation of experience, 49–63
Networks and aging, 176–178
Neural circuits, 25
Neural correlates of consciousness, 56–63, 155
Neural creation of experience, 49–63
Neural firing and experience, 27
Neural impulse, 53–54
Neural mind reading, 56–57
Neural networks, 156–181
Neural processing and perception, 67
Neurons, 25
Neuropsychology, 25, 29

Nobel Prize winners
Adrian, Edgar, 27
Hartline, Keffer, 97
Hubel, David, 60
Ramón y Cajal, Santiago, 27
Sherrington, Charles Scott, 156
Wiesel, Torsten, 60

Occipital lobe damage, 71–72
Octopus, 37–38
Odor perception, unconscious influence, 91
Out-of-body experiences, 21–23

Panpsychism, 35
Paradigm shift, 12–17, 28
cognitive psychology, 12, 16, 28
physics, 12
Parietal lobe damage, 69–71
Parietal reach region, 118
Parsing, 126–127
Philosophical zombies, 44–45
Physiological methods, 24–28, 29t, 30
electroencephalogram, 27, 29, 77–82
event-related potential, 125–126, 135
functional magnetic resonance imaging, 28–30, 56–57, 139
graph theory, 163–165
resting-state fMRI, 159–160
single-neuron recording, 26–27, 55–56, 155
track weighted imaging, 158

Subject Index

Popular culture and the mind, 17–19
Preattentive processing, 75
Prediction, 93–149, 150t, 151–154
  awareness of, 151
  and corollary discharge, 112–119
  and eye movements, 110–114
  and language, 121–128
  machine, mind as, 94
  and memory, 136–140
  and movement, 109–119
  multiple faces of, 149–152
  and music, 126–140
  neural aspects, 104–109
  neurons, 107
  and reaching and grasping, 115–119
  and social behavior, 140–149
  and tickling, 115–116
  time scales, 151
  and touching, 115–119
  and visual perception, 94–115
Prediction error, 105–107
  context based, 138
  creation by artists and composers, 152–154
  in music, 129–130, 152
  syncopation in music, 152
Predictive coding, 105–107
Predictive processing, hierarchical, 108–109
Prefrontal cortex, 105
  and mentalizing, 145
  and mentalizing network, 175
  and mirror network, 144, 145
  and social networks, 175
Priming, 89–91
Problem solving and mind, 4
Processing, bottom-up, 98–99, 104
Processing, top-down, 98, 104

Qualia, 32, 41–49

Readiness potential, 78–82
Receptive fields, visual, 97
Regularities of the environment. *See* Statistical regularities of the environment
Repetition priming, 89–90
Representation, by neurons, 27, 57
Representing networks, 163–165
Resting state fMRI, 159–160
Return to the tonic, 133
Right-hemisphere damage, 73

Salience, 111
Salience network, 161–162
Sally-Anne task, 142, 143
Sentence perception, 125–128
Single-neuron recording, 26, 27, 29, 55–56, 155
Sixties counterculture revolution, 17
Size-weight illusion, 93, 118
Social cognition, 141
  and functional connectivity, 173–176
Social interactions and mind, 4

Social mind wandering, 170
Somato-motor network, 161–162
Somatosensory cortex and mirror network, 148
Speech segmentation, 86–87, 122
Spiritualist approach to the mind, 20
Statistical regularities of the environment. *See also* Knowledge
 and object perception, 99–101
 and language, 86–87
 and predictive coding, 105
 and scanning a scene, 111
Structural connectivity, 157–159
Superior colliculus, 73
Superior temporal sulcus and mirror network, 144–145
Switching between networks, 162–166
Syncopation, 129–132

Task-negative network, 169
Task-unrelated thought, 168–169
Temporal connectivity analysis of functional connectivity, 172
Temporal cortex, 105
Temporal lobe damage, 69–71
Temporal parietal junction (TBJ) and mentalizing, 144, 145
Theory of mind, 142–147
Top-down processing, 98–99, 104
Track-weighted imaging, 158
Transduction, and consciousness, 52–53
Transitional probabilities, 87–88

Unconscious
 brain processes, 80–84
 decision processes, 80–81
 how pathway, 71
 inference, 94–104, 109
 influences on odor perception, 91
 processes, 6, 84–91

Ventral pathway, 68–71
Visual form agnosia, 66–71
Visual kinematics, and movement intention, 147–148
Visual neglect, 73–77
Visual network, 161–162
Visuomotor grip neurons, 118

Wernicke's area, 25–26, 58
What pathway, 68–71
Where/how pathway, 68–71
Word perception, 122–125

Zombies, 44–45, 70